Chromatography of Polymers

Foreword

THE ACS SYMPOSIUM SERIES was first published in 1974 to provide a mechanism for publishing symposia quickly in book form. The purpose of the series is to publish timely, comprehensive books developed from ACS sponsored symposia based on current scientific research. Occasionally, books are developed from symposia sponsored by other organizations when the topic is of keen interest to the chemistry audience.

Before agreeing to publish a book, the proposed table of contents is reviewed for appropriate and comprehensive coverage and for interest to the audience. Some papers may be excluded in order to better focus the book; others may be added to provide comprehensiveness. When appropriate, overview or introductory chapters are added. Drafts of chapters are peer-reviewed prior to final acceptance or rejection, and manuscripts are prepared in camera-ready format.

As a rule, only original research papers and original review papers are included in the volumes. Verbatim reproductions of previously published papers are not accepted.

ACS BOOKS DEPARTMENT

Contents

DETECTION AND DATA ANALYSIS

FIELD FLOW FRACTIONATION AND COUPLED LIQUID CHROMATOGRAPHY METHODS

POLYMER APPLICATIONS

INDEXES

Preface

As we move toward the millennium, the need for a more cost-effective introduction of unique polymer-based products with a high degree of manufacturing quality–consistency has become of paramount importance for industries dependent on polymeric materials as key product components. Competitive pressures also require faster fail-safe product introduction into the marketplace. Coupling the above factors with a constrained set of commercially available polymer building blocks, the emphasis in industry today is on creating polymers with unique structure, topologies, compositions, and morphologies. Sophisticated characterization tools are critical for understanding the nature of complex polymeric materials. Chromatographic characterization of polymers is a key enabler for generating the desired knowledge.

As polymeric materials have become more complex, the characterization and analysis challenges have required the development of new and improved chromatographic methods and data analysis. This has led to an ever-increasing use of hyphenated and multidimensional techniques for analyzing complex polymers. Fortunately, improvements in column technology (resolution and column lifetime), mobile phase delivery systems (accuracy and precision of solvent delivery), and the increasing variety of unique detectors (viscometry, light scattering, MALDI–TOF, IR, evaporative light scattering, diode array UV-vis, etc.) with improved signal to noise ratios, all have enabled improvements and refinements in data analysis and interpretation. Some recent work in this area is discussed in the first section of the book, "Detection and Data Analysis".

The use of field flow fractionation methods, coupled liquid chromatographic methods, and cross-fractionation methods with a variety of detectors involving a range of liquid chromatography modes (size separation, adsorption, desorption, limiting conditions,etc.) are discussed in the second section of this volume, "Field Flow Fractionation and Coupled Liquid Chromatography Methods".

The third section focuses on "Polymer Applications" in which a variety of interesting polymeric materials (polyethylene, metallocene-catalyzed polyolefins, polyesters, polyamides, polyacetylenes, polysaccharides, and PPG-glucan) are subject to analysis using hyphenated and multidimensional techniques to elucidate structure, composition, and branching.

I hope this book will encourage and foster continuing method development and application of hyphenated and multidimensional techniques for characterizing polymers.

Acknowledgments

I am grateful to the authors for their effective oral and written communication and to the reviewers for their critiques and constructive comments. I gratefully acknowledge the ACS Division of Polymeric Materials: Science and Engineering Inc. and ICI PLC for their financial support of the symposium on which this book is based.

THEODORE PROVDER[1]
Consultant
Olmsted Falls, OH 44138

[1]Current address: 26567 Bayfair Drive, Olmsted Falls, OH 44138; e-mail: tprovder@worldnet.-att.net.

DETECTION AND DATA ANALYSIS

Chapter 1

Quantitation in Analysis of Polymers by Multiple Detector SEC

Bernd Trathnigg

Institute of Organic Chemistry, Karl-Franzens-University at Graz, A-8010 Graz, Heinrichstrasse 28, Austria

It is shown, that SEC with dual detection is a powerful tool in the analysis of polymers and also of oligomers. For the latter, one has to take into account the molar mass dependence of response factors, if just one detecto is used. Alternatively, dual detection can be applied also in this case with advantage. Besides the molar mass dependence of response factors, another source of error is the SEC calibration, which may be considerably different not only for different homopolymers, but also for polymer homologous series with the same repeating unit, but different end groups.

Introduction:

In the analysis of polymers by SEC, three transformations are required from chromatographic raw data to molar mass distributions, which may be subject to different sources of error.

- Step 1 (elution time to elution volume) is performed rather easily by using an internal standard for compensation of flow rate variations (assuming the flow rate remaining constant during the run).
- Step 2 (elution volume to molar mass) requires either a calibration function or a molar mass sensitive detector (such as a viscometer or light scattering detector) in addition to the concentration detector(s).
- Step 3 (detector response to weight fraction) is especially important in the case of oligomers and copolymers, where severe errors may result from the assumption of constant response factors over an entire peak.

Consequently, quantitation faces different problems, depending on the nature of the samples and the detectors, which can be applied. In the analysis of oligomers, corrections of response factors for their molar mass dependence have to be made.

In the characterization of copolymers, multiple detection (using combinations of different concentration detectors) is generally inevitable, as it yields additional information on chemical composition and thus allows an accurate quantitation.

In SEC of copolymers or polymer blends, the chemical composition may vary considerably within a peak. In this case, the use of coupled concentration detectors is inevitable.

The concentration detectors most frequently used in SEC of polymers are the UV and the RI detector. Recently, two other detectors have been introduced, which are useful in the analysis of non-UV absorbing polymers: the density detector (according to the mechanical oscillator principle) and the evaporative light scattering detector (ELSD).

The ELSD detects any non-volatile material, but its response depends on various parameters[1],[2], and the nature of these dependencies is rather complex. The UV detector detects UV-absorbing groups in the polymer, which may be the repeating unit, the end groups, or both.

Hence one has to distinguish between polymers, in which the repeating units contain a chromophoric group, and polymers with chromophoric end groups.

In the first case, the response of the UV detector represents the mass, in the second case the number of molecules in a given volume interval. Many chromatographers use this assumption, when they derivatize "non-absorbing" polymers with UV-active reagents. Complications may, however, arise in some mobile phases, as will be discussed later on.

Copolymers: If the response factors of the detectors for the components of the polymer is sufficiently different, the chemical composition of each slice of the polymer peak can be determined from the detector signals. Basically, only very few concentration detectors may be applied: UV-absorbance (UV), refractive index (RI), and density detectors. Infrared (IR) detection suffers from problems with the absorption of the mobile phase, and the evaporative light scattering detector (ELSD) is not suitable for this purpose because of its unclear response to copolymers[3].

For UV-absorbing polymers, a combination of UV absorbance and RI detection is typically used. If the components of the copolymer have different UV-spectra, a diode-array detector can also be applied.

In dual detector SEC there may be different situations in the selection of detectors:

- One component can be detected in UV, the other one does not absorb UV-light. Typical examples are poly(methyl methacrylate-g-ethylene oxide) and poly(methyl methacrylate-g-dimethyl siloxane).

- Both components can be detected in the UV: in the case of poly(methyl methacrylate - b -styrene)[4] the UV spectra of the components are significantly different, and the UV detector can be regarded as selective detector, in the case of poly(methyl methacrylate-b-decyl methacrylate)[5] they are identical, and the UV-absorbance detector has to be regarded as a universal one.
- None of the components can be detected in the UV: this is the case with poly(ethylene oxide-b-propylene oxide) and fatty alcohol ethoxylates (FAE). In the analysis of such samples a combination of density and RI detection can be applied [6,7,8].

The principle of dual detection is rather simple: when a mass m_i of a copolymer, which contains the weight fractions w_A and w_B (= 1 - w_A) of the monomers A and B, is eluted in the slice i (with the volume ΔV) of the peak, the areas $x_{i,j}$ of the slice obtained from both detectors depend on the mass m_i (or the concentration ci =mi/ΔV) of polymer, its composition (w_A), and the corresponding response factors $f_{j,A}$ and $f_{j,B}$, wherein j denotes the individual detectors.

$$x_{i,j} = m_i\left(w_A f_{j,A} + w_B f_{j,B}\right) \qquad \text{equation 1}$$

The weight fractions wA and wB of the monomers can be calculated using equation 2:

$$\frac{1}{w_A} = 1 - \frac{(\frac{x_{i,1}}{x_{i,2}} * f_{2,A} - f_{1,A})}{(\frac{x_{i,1}}{x_{i,2}} * f_{2,B} - f_{1,B})} \qquad \text{equation 2}$$

and therefrom the mass of polymer in the corresponding interval

$$m_i = \frac{x_{i,1}}{w_A * (f_{1,A} - f_{1,B}) + f_{1,B}} \qquad \text{equation 3}$$

Once the amount of polymer in an interval and its chemical composition are known, one may calculate the corresponding molar mass (transformation 2). In the case of copolymers, the molar mass M_C of the copolymer can obtained by interpolation between the calibration lines of the homopolymers[9] (which may be considerably different)

$$\ln M_C = \ln M_B + w_A * (\ln M_A - \ln M_B) \qquad \text{equation 4}$$

wherein M_A and M_B are the molar masses of the homopolymers, which would elute in this slice of the peak (at the corresponding elution volume V_e)[10].

There is , however, still a chemical polydispersity in each slice, which means, that M_C is just the average molar mass, since w_A is also an average composition.

Obviously, the precision as well as the accuracy of the results obtained by this technique will depend on the individual response factors. Thus, it is important to find an appropriate combination of detectors and mobile phase in order to obtain reliable results. This can, however, sometimes require a lot of experiments, which means also a lot of trial and error.

In a recent study[11], we have shown, that a simple simulation procedure can considerably reduce the time required for the optimization of such a method. Basically, one has to determine the response factors of both detectors for the homopolymers, and calculate the peak areas, which would result from different amounts of sample (say 1 - 10 µg) with compositions between 0 and 100 %. Then the smaller area is increased by 1 digit (thus simulating a baseline uncertainty), and the composition is again calculated for the new peak areas. The error in composition for sample sizes of 1- 5 µg is a good criterion for evaluating the suitability of a mobile phase with a given detector combination.

In Figure 1, such a plot is shown for the system density + RI detector with chloroform as mobile phase applied to the analysis copolymers of ethylene oxide (EO) and propylene oxide (PO).

Even though this is not the most favourite case, good results can be obtained, as can be seen from Figures 2 - 4, which show a chromatogram of an EO-PO block copolymer and the MMD and chemical composition calculated therefrom. This sample obviously contains a fraction of PPG, as could be proven by two-dimensional LC[12].

While the MMD in Figure 3 was obtained with the calibration function for PEG, the molar mass for each interval was calculated using equation 4 to yield the MMD shown in Figure 4. As can be seen, the molar mass averages thus obtained are considerably different !

It must be mentioned, that also the molar mass dependence of response factors has been accounted for in these results, as will be discussed in the following section.

Oligomers: In the analysis of oligomers, additional problems arise from the fact, that the response of most detectors depends more or less strongly on the molar mass of the samples.

The UV detector detects UV-absorbing groups in the polymer, which may be the repeating unit, the end groups, or both.

Hence one has to distiguish between polymers, in which the repeating units contain a chromophoric group, and polymers with chromophoric end groups.

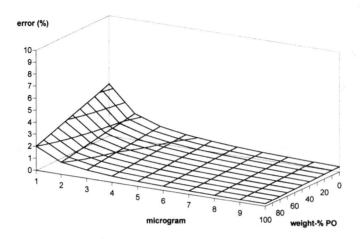

Figure 1: Simulation of dual detector SEC of EO-PO copolymers in chloroform with coupled density and RI detection

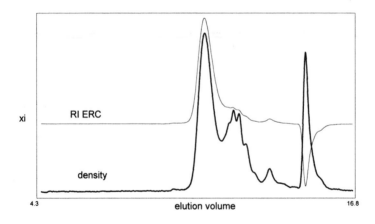

Figure 2: Chromatogram of an EO-PO block copolymer, as obtained by SEC in chloroform with coupled density and RI detection (Plgel 5µm 100 Å, 600 x 7.8 mm)

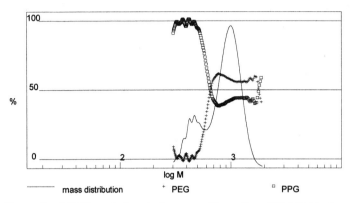

Figure 3: MMD and chemical composition of an EO-PO block co-polymer (Figure 2), as determined by SEC in chloroform with density and RI detection. Calibration: PEG,

Molar mass averages: $M_w = 801$ $M_n = 680$ $M_w/M_n = 1.214$

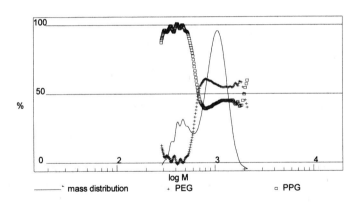

Figure 4: MMD and chemical composition of an EO-PO block copolymer (Figure 2), as determined by SEC in chloroform with density and RI detection. Calibration: PEG+ PPG

Molar mass averages: $M_w = 845$ $M_n = 696$ $M_w/M_n = 1.214$

In the first case, the response of the UV detector represents the mass, in the second case the number of molecules in a given volume interval. Many chromatographers use this assumption, when they derivatize "non-absorbing" polymers with UV-active reagents. Complications may, however, arise in some mobile phases, as will be discussed later on.

RI and density detector measure a property of the entire eluate, which is related to a specific propertiy of the sample (the refractive index increment or the apparent specific volume, respectively).

It is a well known fact[13], that specific properties are related to molar mass

$$x_i = x_\infty + \frac{K}{M_i}$$

Equation 5

wherein x_i is the property of a polymer with the molar mass M_i, x_∞ is the property of a polymer with infinite (or at least very high) molar mass, and K is a constant reflecting the influence of the end groups[14,15,16]. A similar relation describes the molar mass dependence of response factors for RI and density detection.

$$f_i = f_\infty + \frac{K}{M_i}$$

Equation 6

In the case of the evaporative light scattering detector (ELSD) no such simple relation exists, and the (more volatile) lower oligomers can be lost at higher evaporator temperatures[17].

In a plot of the response factors f_i of polymer homologous series (with defined end groups) as a function of $1/M_i$ straight lines will be obtained. Their intercept f_∞ can be considered as the response factor of a polymer with infinite molar mass, or the response factor of the repeating unit, the slope K represents the influence of the end groups[18]. The magnitude of K, determines the molar mass range, above which response factors can be considered as constant, which is often the case only at molar masses of several thousands. Neglecting this dependence can lead to severe errors, as has been shown[19] in the analysis of ethoxylated fatty alcohols (FAE) by SEC in chloroform. Once f_∞ and K have been determined, the correct response factors for each fraction with the molar mass M_i (which is obtained from the SEC calibration).

In a previous paper[20], three methods were described for the determination of f_∞ and K:

1. If a sufficient number of monodisperse oligomers is available, as is the case with PEG and PPG, one may determine the individual response factors and calculate f_∞ and K by linear regression, as is shown in Fig.5. As has been shown previously, this approach is also possible with fatty alcohol ethoxylates (FAE).

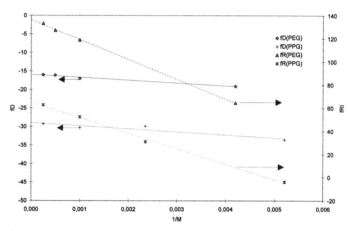

Figure 5: Response factors of polyethyleneglycol (PEG) and polypropyleneglycol (PPG) in chloroform for density and RI detection

2. If only one monodisperse oligomer is available (such as the fatty alcohol for some FAE) and f_∞ can be determined otherwise (either as
above or as a good approximation - from a sufficiently high molecular
sample), a two-point calibration may work quite well.
3. If no monodisperse oligomer is available, but f_∞ is known, an iteration
procedure may be used, which is shown schematically in Fig. 6.

The latter procedures can also be applied to macromonomers, for which no
monodisperse oligomers are available.

Obviously, the same approach should be applicable to UV detection of
polymers with UV-absorbing end groups.

Figures 7 - 9 show the response factors of density, RI and UV detection (at
260 nm) as a function of 1/M.

For all polymer homologous series with the same repeating unit the same
intercept must be expected, while the different end groups should result in
a different slope. Linear regression was used for PEG, PEG-mono- and -
dimethacrylates, and methyl-PEG-monomethacrylates; the iteration procedure (approach 3) was applied in all cases, and the slopes thus obtained
were in good agreement with those from linear regression.

A strange phenomenon is observed with the UV detector. while the oli-
gomers with methacrylic end groups follow equation 6 (as expected),
PEGs of a molar mass larger than a few hundreds produce large peaks in
UV at a wavelength, where no adsorption could be expected (260 nm)!
This is especially bad news for anybody, who tries to determine the func-
tionality of such products after derivatization.

The same behaviour can also be observed with fatty alcohol ethoxylates
(FAE), as can be seen in Fig.10.

The reason for these large signals is not yet clear. Because of these complications, UV detection was not applied in chloroform any more.

While the response factors of the density detector are always negative in
chloroform (regardless the nature of the end groups), those of the RI detector change their sign a lower molar mass, as can also be seen from
Fig.11, which shows a chromatogram of a PEG-200-monomethacrylate,
obtained by SEC in CHCL3 with density and RI detection.

Obviously, the lower oligomers have a negative sign in RI detection, while
no problems occur in density detection. From the density data, the MMD
was calculated using equation 6. Obviously, such a correction cannot be
applied for the refractive index trace.

Alternatively, functional oligomers may be regarded as block copolymers,
consisting of one block without end groups, and another one: the end
groups; in this case this means PEG and the methyl methacrylate (MMA),
respectively. Using equations 2 and 3, the MMD shown in Figure 12 was
calculated, which agreed perfectly with that from the density data and
equation 6.

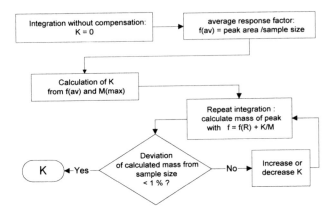

Figure 6:   Iteration procedure for determination of K for polydisperse oligomers

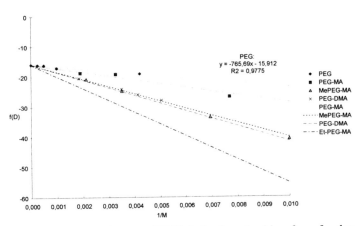

Figure 7: Response factors of PEG derivatives in chloroform for density detection

12

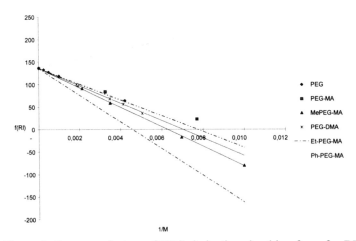

Figure 8: Response factors of PEG derivatives in chloroform for RI detection

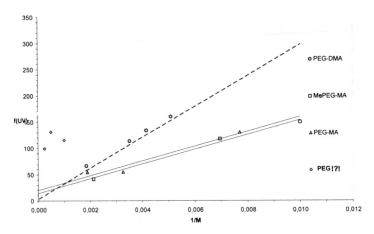

Figure 9: Response factors of PEG derivatives in chloroform for UV detection at 260 nm

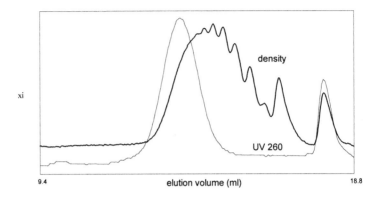

Figure 10: Chromatogramm of a fatty alcohol ethoxylate based on 1-dodecanol (Brij 30), as obtained with density and UV detection (260 nm)

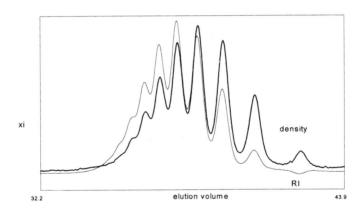

Figure 11: Chromatogram of a macromonomer, methoxy-PEG-200-methacrylate, as obtained on a set of 5 columns Phenogel 5μm (300x7.8 mm each, 2 x 500Å + 3 x 100 Å) in chloroform with coupled density and RI detection

14

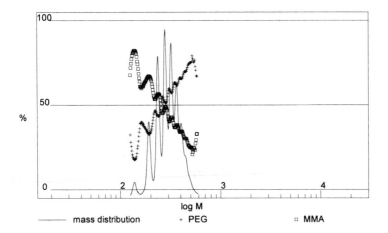

Figure 12: MMD and chemical composition of the macromonomer in Figure 11, as determined from dual detection

Figure 13 shows the effect of different corrections on the molar mass averages for this sample. Obviously, the agreement between the values from the density data using eqn. 6 and those from dual detection is very good.

Additionally, it becomes clear, that there is a big difference in the values calculated with the PEG calibration and those obtained with the correct calibration for this homologous series.

As a test for the performance of the dual detection method, we have analyzed methoxyethyl methacrylate, which represents the oligomer 1 (containing 1 MMA + 1 EO). The result is shown in Figure 14: A nice horizontal line if found for the composition, which indicates, that as well the delay volume between the detectors is correct, and that peak dispersion between the detectors is negligible. The EO content was found to be 28.6%, which agrees quite well with the theoretical one (30.6).

In Figures 15 and 16, the same comparison is given for samples with different end groups and different molar mass. As before, the correction of density data using equation 6 and the dual detector method agree very well, while the correction of RI data is not as efficient, if it is possible at all.

Experimental

The polyethylene glycols used as calibration standards were purchased from Polymer Laboratories (Church Stretton, Shropshire, UK), the macro-monomers MePEG-200-MA and MePEG-400-MA from Polysciences (Warrington, PA), all other samples from FLUKA (Buchs, Switzerland),

All measurements were performed on a modular SEC system comprising of a Gynkotek 300C pump equipped with a VICI injector (sample loop 100 µl), two column selection valves Rheodyne 7060, a density detection system DDS 70 (Chromtech, Graz, Austria) coupled with an ERC 7512 RI detector or a JASCO 875 UV absorbance detector.

Data acquisition and processing was performed using the software CHROMA, which is part of the DDS 70.

The following columns were used: Phenogel M (5 µm), 600x7.6 mm, PL Microgel M (10 µm), 600x7.6 mm, a set of four columns Phenogel (5µm), 2x500 and 2x100 Å, 300x7.6 mm each, and a set of two columns PL Microgel (10µm), $10^3 + 10^4$ Å, 300x7.6 mm each.

All measurements were performed at a flow-rate of 1.00 ml/min and a column temperature of 30.0°C. Sample concentrations were 3.0 - 10.0 g/l.

The chloroform used in this study was HPLC grade and stabilized with 2-methyl-butene (Mallinckrodt).

Acknowledgement

Financial support by the Austrian Academy of Sciences (Grant OWP-53) is gratefully acknowledged.

16

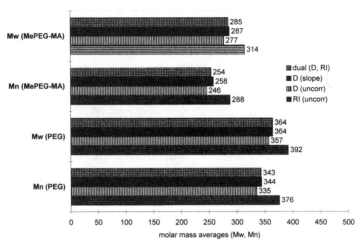

Figure 13: Molar mass averages of methoxy-PEG-200-methacrylate, as obtained on Plgel 5µm 100 Å, 600 x 7.8 mm in chloroform with coupled density and RI detection with and without correction

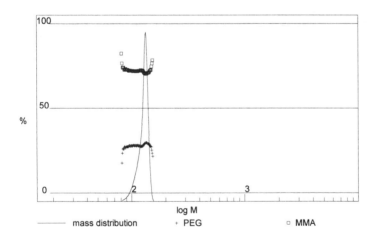

Figure 14: MMD and chemical composition of the methoxyethyl methacrylate, as determined from dual detection.

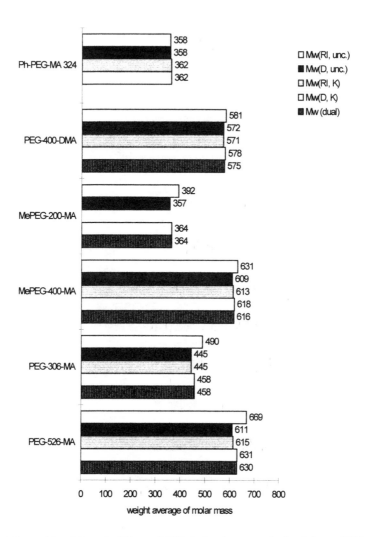

Figure 15: M$_w$ of different PEG derivatives, as obtained from SEC in chloroform with density and RI detection, with and without correction of detector response

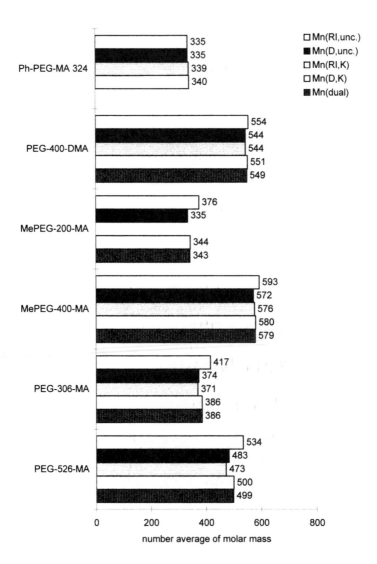

Figure 16: M_n of different PEG derivatives, as obtained from SEC in chloroform with density and RI detection, with and without correction of detector response

References:

[1]   Vandermeeren, P.; Vanderdeelen, J.; Baert, L.; *Anal Chem* **1992,** *64*, 1056.

[2]   Dreux,M; Lafosse, M; Morin-Allory,L. *LC-GC Int.* **1996,** *9*, 148.

[3]   Chiantore, O.; *Ind. & Eng. Chem. Res.*, in press

[4]   Pasch,H.; Gallot, Y.; Trathnigg B.: *Polymer* **1993,** *34*, 4985.

[5]   Pasch, H.; Augenstein, M.; Trathnigg, B.: *Macromol. Chem. Phys.* **1994,** *195*, 743.

[6]   Trathnigg, B., *J.Liq.Chromatogr. 1990, 13,* 1731.

[7]   Trathnigg, B.; *J.Chromatogr.* **1991,** *552*, 505.

[8]   Trathnigg, B.; Yan, X.; *Chromatographia* **1992,** *33*, 467.

[9]   Runyon, J.R.; Barnes, D.E.; Rudel, J.F.; Tung, L.H.; *J. Appl. Polym. Sci.* **1969,** *13*, 2359.

[10]   Tung, L.H.; *J. Appl. Polym. Sci.* **1979,** *24*, 953

[11]   Trathnigg, B; Feichtenhofer, S.; Kollroser, M.; *J.Chromatogr. A,* in press

[12]   Trathnigg, B.; Kollroser, M.; in preparation

[13]   Candau, F.; Francois, J.; Benoit, H.; *Polymer* **1974,** *15*, 626.

[14]   Schulz, G.V.; Hoffmann, M.; *Makromol.Chem.* **1957,** *23*, 220.

[15]   Kobataka Y.; Inagaki, H.; *Makromol.Chem.* **1960,** *40*, 118.

[16]   Géczy, I.; *Tenside Deterg.* **1972,** *9*, 117.

[17]   Trathnigg, B.; Kollroser, M.; *J.Chromatogr.A* **1997,**768, 223.

[18]   Cheng, R.; Yan, X.; *Acta Polym. Sin.* **1989,** *12*, 647.

[19]   Trathnigg, B.; Thamer, D.; Yan, X.; Maier, B.;.Holzbauer, H-R.; Much, H.; *J.Chromatogr. A* **1993,** *657*, 365

[20]   Trathnigg, B.; Yan, X.; *J.Appl.Polym.Sci.: Appl.Polym.Symp.* **1993,** *52*, 193

Chapter 2

Use of Multidetector SEC for Determining Local Polydispersity

T. H. Mourey[1], K. A. Vu[1], and S. T. Balke[2]

[1]Imaging Research and Advanced Development, Eastman Kodak Company, Rochester, NY 14650–2136
[2]Department of Chemical Engineering and Applied Chemistry, University of Toronto, Toronto, Ontario M5S 3E5, Canada

Local polydispersity refers to the presence of more than one type of polymer molecule at a particular retention volume in size exclusion chromatography (SEC). It can be a serious source of error in SEC analysis. In this work we show that conventional multidetector SEC data can readily be used to detect the presence of local polydispersity. The method utilizes molecular weight calibration curves obtained from viscometer and light-scattering detectors combined with a comparison of a differential refractive index (DRI) reconstructed chromatogram and the experimental one. Origins of local polydispersity were classified and shown in a schematic diagram together with the resulting types of calibration curves and reconstructed chromatograms. The independence of the reconstructed chromatogram to variation in the specific refractive index increment across the chromatogram greatly enhanced the effectiveness of the method.

Local polydispersity can be a serious source of error in SEC analysis but is generally assumed negligible. It occurs when different types of polymer molecules elute at the same SEC retention volume. Axial dispersion is one source of local polydispersity. However, local polydispersity can occur even when resolution is perfect. For a linear copolymer or blends of polymers, for example, different combinations of molecular weight and composition can result in the same molecular size and hence the same retention volume.

If local polydispersity is present and is significantly affecting conclusions from the analysis, then the conventional SEC analysis must be supplemented with other experimental methods. For example, chromatographic cross fractionation *(1)* or a new technique utilizing special SEC sample preparation methods *(2)* can be used. The objective of this work is to define the capability of conventional multidetector SEC analysis to detect local polydispersity in a particular sample and thus determine when such additional experimental work is necessary.

Theory

Local Concentration From the DRI Chromatogram. The need to accurately determine local concentration is often a significant difficulty in determining the presence of local polydispersity. The local concentration (the total polymer concentration at each retention volume, c_i) is used to calculate local weight average molecular weight from light-scattering detectors, local intrinsic viscosity from viscometer detectors, and the value plotted on the vertical axis of the molecular weight distribution for all detectors. The true value of local concentration, $c_{i, true}$, can be obtained from the DRI chromatogram heights, W_i, at each retention volume, v_i, only if the specific refractive index increments $(dn/dc)_{ij}$ of the j different types of molecules at retention volume v_i, and their corresponding component DRI chromatogram height, W_{ij} are known:

$$c_{i,true} = \frac{\dfrac{W_{i1}}{\beta\left(\dfrac{dn}{dc}\right)_{i1}} + \dfrac{W_{i2}}{\beta\left(\dfrac{dn}{dc}\right)_{i2}} + \dots \dfrac{W_{ij}}{\beta\left(\dfrac{dn}{dc}\right)_{ij}}}{\sum\limits_{i=1}^{\infty}\left[\dfrac{W_{i1}}{\beta\left(\dfrac{dn}{dc}\right)_{i1}} + \dfrac{W_{i2}}{\beta\left(\dfrac{dn}{dc}\right)_{i2}} + \dots \dfrac{W_{ij}}{\beta\left(\dfrac{dn}{dc}\right)_{ij}}\right]\Delta v_i} \times m \tag{1}$$

where β is the DRI detector response constant. Normally, the contribution of individual molecules to the DRI detector response is not known, and we assume that the specific refractive index increment is the same for all molecules j within a slice and for all retention volumes, i, across the size distribution. Then, the DRI chromatogram heights and the DRI chromatogram area are used to calculate an "apparent" local concentration:

$$c_{i,app} = \frac{W_i}{\sum\limits_{i=1}^{\infty} W_i \Delta v_i} m \tag{2}$$

However, equation 2 is only valid if the concentration detector response is proportional to the concentration of polymer at each retention volume and if this "detector response constant" does not change across the chromatogram. A major factor confounding determination of local polydispersity is the variation of the concentration detector response constant (or more generally, the relationship between concentration and concentration detector response) across the chromatogram. For example, even with no significant local polydispersity, the detector response for a copolymer from a DRI detector generally varies across the chromatogram because the local average composition of the molecules varies with molecular size. This situation is well known and the use of a DRI together with a UV spectrophotometer has often been used to obtain the average concentration of each of the two monomer species present in the copolymer at each retention volume (i.e., the local average composition) (4).

Thus, inaccuracy in the calculated local concentration values from equation 2 is the first obstacle to determining the effect of local polydispersity. Variation of the detector response constant across the chromatogram can be assessed by comparing the normalized chromatograms of the DRI and UV detectors. Superposition of the two chromatograms is an indication that the detector response constant does not vary. However, this is not always true. For example, it is possible that one component is present that is not detected by either detector (the detector response constant for one component is zero).

Local Number Average Molecular Weight From the Viscometry Detector. From the work of Hamielec and Ouano *(3)* we know that, if local polydispersity is present, the effective hydrodynamic volume in SEC becomes the product of intrinsic viscosity and local number average molecular weight. This means that, when the universal calibration curve is used with a viscometer and concentration detector, the resulting molecular weight at retention volume i is the local number average value:

$$M_{n_{i,true}} = \frac{J_i c_{i,true}}{\eta_{sp,i}} \tag{3}$$

The use of apparent local concentration results in apparent local number average molecular weights that may or may not be correct, depending on the validity of equation 2:

$$M_{n_{i,app}} = \frac{J_i c_{i,app}}{\eta_{sp,i}} \tag{4}$$

Local Weight Average Molecular Weight From the Light-Scattering Detector. A less well-recognized consequence of local polydispersity is that the diversity of differential refractive index increment (dn/dc) values at a particular retention volume can cause large errors in the local weight average molecular weight estimated from light scattering. When variation of dn/dc is present at a particular retention volume, the light-scattering response is the sum of the individual light- scattering responses for each type of molecule present,

$$M_{w_{i,true}} = \frac{1}{\alpha c_{i,true}} \sum_{j=1}^{k} \frac{R_{\theta_{ij}}}{P(\theta)_{ij}\left(\frac{dn}{dc}\right)_{ij}^2} \tag{5}$$

The usual equation used to determine local weight average molecular weight from the apparent local concentration is

$$M_{w_{i,app}} = \frac{R_{\theta_i}}{\alpha P(\theta)_i c_{i,app}\left(\frac{dn}{dc}\right)^2} \tag{6}$$

where α is an optical constant and the specific refractive index increment is an average value for the whole sample, obtained from the integrated DRI chromatogram:

$$\overline{\frac{dn}{dc}} = \frac{1}{\beta m} \sum_{i=1}^{\infty} W_i \Delta v_i \qquad (7)$$

Equations 5 and 6 can provide dramatically different values of $M_{w,i}$ and are particularly sensitive to the values of dn/dc used.

Reconstructed DRI Chromatograms. Recently, we reported a method for detecting local polydispersity by reconstructing the DRI chromatogram from light-scattering and viscometry detector signals, through the universal calibration curve *(5)* If no local polydispersity is present, then local molecular weight averages calculated from equations 4 and 6 are equal. Equating and rearranging the two expressions and using the relationship between DRI chromatogram height and local concentration:

$$c_i = \frac{W_i}{\beta \left[\dfrac{dn}{dc} \right]_i} \qquad (8)$$

permits reconstruction of the DRI chromatogram that does not depend on the specific refractive index increment,

$$W_{i,\text{reconstructed}} = \beta \left(\frac{\eta_{sp_i} R_{\theta_i}}{\alpha P(\theta)_i J_i} \right)^{\frac{1}{2}} \qquad (9)$$

where hydrodynamic volume, J_i, is obtained at each retention volume from the universal calibration curve. Equation 9 reconstructs the DRI chromatogram only if: (a) there is no local molecular weight polydispersity, (b) there is no local polydispersity of any type that causes a variety of dn/dc or $P(\theta)$ values to be present at v_i, and (c) universal calibration applies. We have recently shown both theoretically and experimentally that the reconstructed chromatogram is unaffected by the variation of dn/dc across the chromatogram and that the method is capable of detecting local polydispersity in either molecular weight or chemical composition, provided the latter results in a variety of dn/dc values at each retention volume *(5)*.

The method most closely resembling this one in the published literature is the recent work of Brun *(6)*. He employs Equation 9 written explicitly in terms of a "reconstructed", or as he terms it, an "observed" value of hydrodynamic volume, J_i. He uses the experimental value of the DRI detector response, W_i, in Equation 9 rather than the experimental value of J_i as we do. His analysis focuses upon the difference between the reconstructed J_i values and the experimental values of J_i. Although, as Brun points out, the local polydispersity can be discerned by examining this difference, his emphasis is more on factors such as adsorption and concentration effects which cause violation of the universal calibration curve assumption. Our approach assumes that universal calibration is valid and focuses upon the various types of local polydispersity.

The experimentally accessible quantities are the apparent values defined by equations 2, 4 and 6, the average specific refractive index increment (equation 7), and the reconstructed chromatogram (equation 9). The reconstructed chromatogram is particularly well suited for detecting local polydispersity because it is independent of the concentration detector as well as the average specific refractive index increment.

Thus, it is sensitive only to variety in molecules at a specific retention volume. Any significant difference between the reconstructed chromatogram and the actual DRI chromatogram indicates local polydispersity. However, whether the local polydispersity originates from a significant difference in molecular weights or composition or both is not evident. To obtain more information on the origin of the local polydispersity we use molecular weight calibration curves along with a comparison of reconstructed and experimental DRI chromatograms.

Unlike reconstructed DRI chromatogram heights, local molecular weight averages from viscometry and from light scattering rely on apparent concentrations that are sensitive to variation in dn/dc across the chromatogram. Also, number average molecular weight from viscometry is directly proportional to the apparent local concentration (equation 4) while the local weight average molecular weight from light scattering is inversely proportional to the apparent local concentration (equation 6). This can cause noticeable differences between the two local molecular weight averages when the apparent local concentrations are incorrect. Furthermore, the disparity between the two local averages can be further exaggerated by the large influence of the average refractive index increment on the local weight average from light scattering (equation 6).

Here we show the initial steps in developing a method for detecting local polydispersity and for learning about its origin from multidetector SEC alone. In this work, we utilize binary polymer blends so that the local polydispersity is known. As mentioned above, the method is based upon using both molecular weight calibration curves and a comparison of reconstructed and experimental DRI chromatograms. The end result is a flow chart showing how different types of variability in local concentration and dn/dc can be discerned by the method. This flow chart is shown in Figure 1. Each branch in this chart will be individually examined in the following sections.

Experimental

The SEC detectors, arranged in series, are a 757 Spectroflow spectrophotometric detector (UV), a Precision Detectors PD2000 light-scattering (LS) detector operating at 15 and 90 degrees, a Viscotek H502A differential viscometer (DV), and a Waters 411 differential refractive index (DRI) detector. The columns, LS, DV, and DRI detector temperatures were maintained at 35°C. The eluent was uninhibited tetrahydrofuran, nominally delivered at a flow rate of 1.0 mL/min. Flow rate corrections were made using 0.2% acetone added to the sample solvent as a flow marker. Columns were three Polymer Laboratories PLgel mixed-C, 7.5 x 300 mm. Polymer blends at a 1:1 weight ratio were injected at a total concentration of 1.5 mg/mL in an injection volume of 100 μL.

To supplement the experimental results computer simulations were also run. The software developed used the chromatograms of separately injected blend components to generate the needed information for the polymer blend. It permitted dn/dc values and other parameters to be arbitrarily varied so that the method of detecting local polydispersity could be tested on an expanded variety of cases.

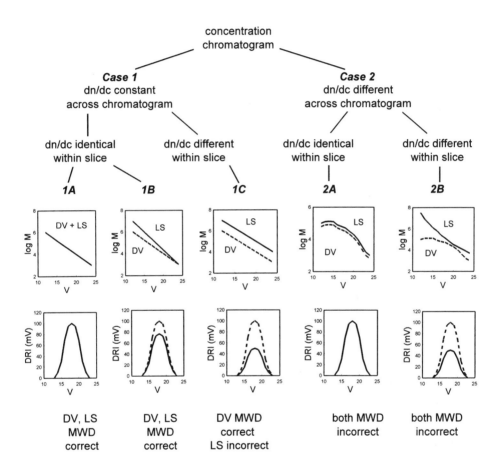

Figure 1.   Local polydispersity flow chart.

Results and Discussion

Case 1. Average dn/dc at Each Retention Volume, Identical from One Retention Volume to Another, Across the Chromatogram.

1A (Figure 1): No Local Polydispersity. All molecules have identical molecular weight, dn/dc and $P(\theta)$ at each retention volume. Since the detector response constant, β, in equation 8 does not vary across the chromatogram, then equation 2 is valid. Then accurate local $M_{n,i}$ values can be obtained from equation 4 using a viscometer detector and plotted as a molecular weight calibration curve. If $M_{w,i}$ is obtained from equation 6 using the light-scattering detector and the resulting molecular weight calibration curve superimposes on the local $M_{n,i}$ curve, this indicates no local polydispersity ($M_{n,i} = M_{w,i} = M_i$). It also shows that the interdetector volumes between the detectors are correctly defined and that axial dispersion effects are negligible. Thus, injection of a linear homopolymer provides a "base case" that should be used to ensure that the system has been properly set up before proceeding further. In our case, polystyrene was used for this purpose and superposition of the calibration curves as well as the reconstructed and experimental DRI chromatograms was obtained. Molecular weight distributions from viscometry and light-scattering are identical as expected.

1B (Figure 1): Local Molecular Weight Polydispersity Present. Examples include local polydispersity arising from significant mixing of molecules caused by axial dispersion, or a variety of molecules with the same hydrodynamic volume but different molar masses. The latter example is shown in Figure 2 for a mixture of linear low molecular weight and lightly branched higher molecular weight polyester. Here, dn/dc values are identical across the chromatogram but a mixture of branched and linear molecules exists at many retention volumes. Since equations 2, 4, and 6 are valid, the $M_{w,i}$ and $M_{n,i}$ results are accurate. The small difference between the two calibration curves shows that only a low level of local molecular weight polydispersity (as measured by the two averages) is present. This is confirmed by comparison of the reconstructed DRI chromatogram to the experimental chromatogram for the blend (Figure 3). Only small differences are observed. Thus, we recognize samples of this class from small differences in both the calibration curves and the DRI chromatograms. Molecular weight distributions from light scattering and viscometry detection are expected to be correct although not identical.

1C (Figure 1): Local Composition (dn/dc) Polydispersity Present. To illustrate the effect of local variablity in dn/dc on the two calibration curves, a 1:1 blend of polystyrene and poly(dimethylsiloxane) (PDMS) where the two polymers had identical molecular size distributions but different dn/dc values were analyzed. The dn/dc of polystyrene is 0.180 while that of PDMS is 0.003. Thus, although the average dn/dc value did not vary across the chromatogram, molecules at any particular retention volume had a large difference in dn/dc. Figure 4 shows the calibration curves obtained from the DV detector (equation 4) and the light-scattering detector (equation 6). Some difference between the curves is expected because PS

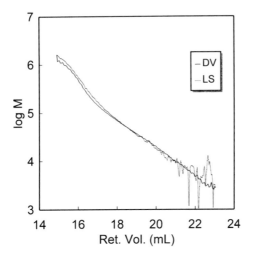

Figure 2.   Calibrations curves from light-scattering (LS) and differential viscometry (DV) detection for 1:1 blend of linear and branched polyester (Case 1B).

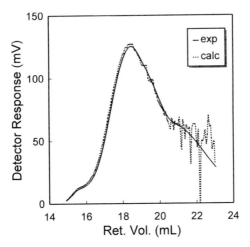

Figure 3.   Reconstructed (calc) and experimental (exp) DRI chromatograms for 1:1 blend of linear and branched polyester (Case 1B).

28

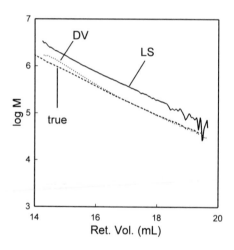

Figure 4.   Calibration curves from LS and DV detection for 1:1 blend of PS and PDMS (Case 1C).

and PDMS have different Mark Houwink constants. However, the difference shown is inaccurate because while the $M_{n,i}$ curve from the DV detctor is valid, the $M_{w,i}$ is not. That is, equation 5 rather than equation 6 should have been used to calculate $M_{w,i}$, but this would not be feasible for an unknown sample.

It was shown previously that the reconstructed chromatogram is considerably larger than the experimental DRI chromatogram *(5)*. This lack of superposition combined with nearly parallel log M calibration curves obtained from light scattering and viscometry detection provide a unique signature for this class of local polydispersity. The molecular weight distribution obtained from viscometry detection is more accurate while the distribution obtained from light scattering is not.

Case 2. Average dn/dc at Each Retention Volume Varies from One Retention Volume to Another, Across the Chromatogram.

2A (Figure 1): All Molecules at a Particular Retention Volume Have the Same dn/dc. We used the computer simulation to demonstrate this case. The DRI chromatograms for the individual branched and linear components of Case 1 (Figure 5) provided the needed arbitrary chromatograms. The dn/dc of component 1 (the chromatogram of the linear polymer) was set to 0.123 and the dn/dc of component 2 (the branched component) was set to 0.003. The resulting blend has a variation in local average dn/dc across the distribution (Figure 6) as the ratio of the two components varied with retention volume. However, by using the computer simulation, the dn/dc of every molecule at a particular retention volume was set to this local average value for that particular retention volume. The Rayleigh scattering is then calculated at each retention volume and the local molecular weights are calculated from the apparent concentrations to generate the calibration curves for viscometry and light-scattering, shown in Figure 7. The apparent calibration curves are similar to each other, but clearly different from the true calibration curve because equation 2 (local concentration) does not apply. The reconstructed calibration curve is sensitive only to local polydispersity and not to variation of dn/dc across the chromatogram, and superposition with the experimental DRI curve is obtained (Figure 8). The molecular weight distributions obtained from both molecular-weight-sensitive detectors are inaccurate.

2B (Figure 1): Molecules with Different dn/dc Values are Present at Each Retention Volume and the Average dn/dc Varies Across the Chromatogram. Again the computer simulation was used. The same chromatograms were used as in Case 2A, except now molecules exist at each retention volume with dn/dc = 0.123 (component 1) and dn/dc = 0.003 (component 2). That is, this time the molecules at a particular retention volume were allowed to retain their respective dn/dc values (i.e. unlike Case 2A, where they were each assigned the average dn/dc value). The calibration curves from viscometry and light scattering are different from each other and do not agree with the true calibration curves (Figure 9). The reconstructed DRI chromatogram does not superimpose on the experimental chromatogram (Figure 10) because of local polydispersity that gives

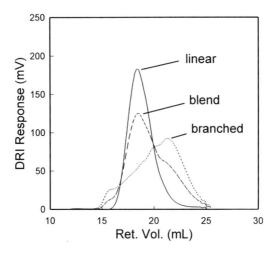

Figure 5. Experimental DRI chromatograms for linear and branched polyester, and 1:1 blend.

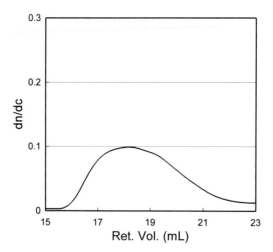

Figure 6. Local average specific refractive index increment variation across chromatogram from addition of chromatograms in Figure 5.

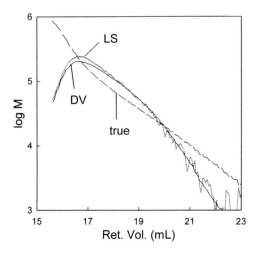

Figure 7.   Apparent calibration curves from DV and LS detection from simulation for dn/dc different across the chromatogram but identical within a slice, and true calibration curve (Case 2A).

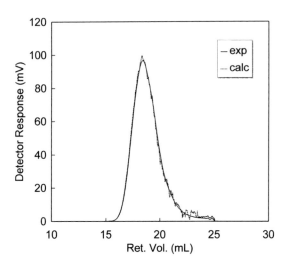

Figure 8.   Reconstructed and experimental DRI chromatograms simulation for dn/dc different across the chromatogram but identical within a slice (Case 2A). Reconstructed curve is the noiser of the two.

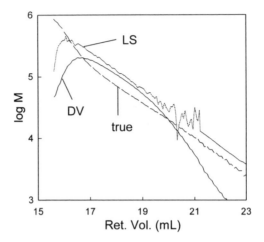

Figure 9.   Apparent calibration curves from DV and LS detection from simulation for dn/dc different across the chromatogram and varying within a slice, and true calibration curve (Case 2B).

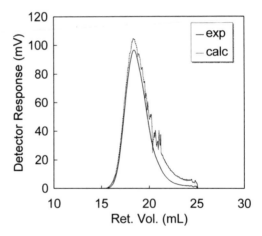

Figure 10.   Reconstructed and experimental DRI chromatograms simulation for dn/dc different across the chromatogram and varying within a slice (Case 2B).

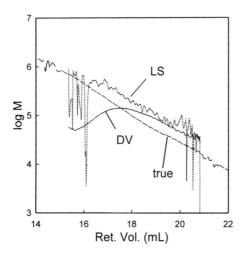

Figure 11.   Calibration curves for 1:1 blend of PMMA and PDMS (Case 2B).

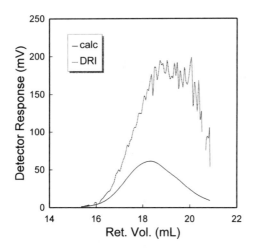

Figure 12.   Reconstructed and experimental DRI chromatograms for 1:1 blend of PMMA and PDMS (Case 2B).

rise to a variety in dn/dc at each retention volume. Molecular weight distributions obtained from both viscometry and light-scattering detection are incorrect.

Simulation results were accompanied by experimental results for a blend of poly(methylmethacrylate) (M_w = 80,000, dn/dc = 0.086) and PDMS (M_w = 265,000, dn/dc = 0.003). Unusual calibration curve shapes are obtained and neither agrees with the true distribution (Figure 11) nor does the reconstructed chromatogram superimpose on the experimental DRI chromatogram (Figure 12).

Conclusions

Molecular weight calibration curves in multidetector SEC do communicate the presence of local polydispersity. However, interpreting their message is made particularly difficult by dn/dc variability amongst the molecules in a complex polymer (such as a copolymer or a branched polymer). The method presented here utilizes molecular weight calibration curves obtained from viscometer and light-scattering detectors combined with a comparison of a DRI reconstructed chromatogram and the experimental one. Origins of local polydispersity were classified by emphasizing the type of dn/dc variability that can be present. They were shown in a schematic diagram together with the resulting types of calibration curves and reconstructed chromatograms. The independence of the reconstructed chromatogram to variation in dn/dc across the chromatogram greatly enhanced the effectiveness of the method.

Literature Cited

1. Glöckner, G. *Gradient HPLC of Copolymers and Chromatographic Cross-Fractionation*, Springer-Verlag: Berlin, 1991.
2. Thitiratsakul, R.; Balke, S. T.; Mourey, T. H.; Schunk, T. C., "Detecting Perfect Resolution Local Polydispersity in Size Exclusion Chromatography", *Int. J. Polym. Anal. Char.*, accepted for publication
3. Hamielec, A. E.; Ouano, A. C. *J. Liq. Chromatogr.* **1978**, *1*, 111.
4. Dawkins, J. V. *Adv. Chem. Ser.* **247**, 197 (1995).
5. Mourey, T. H.; Balke, S. T. *J. Appl. Polym. Sci.*, in press.
6. Y. Brun, *International GPC Symposium '96 Symposium Proceedings*, Waters Corporation, 1996, pp. 116-153

Chapter 3

The Chevron Approach to GPC Axial Dispersion Correction

Wallace W. Yau

Chevron Chemical Company LLC, Kingwood Technical Center, 1862 Kingwood Drive, Kingwood, TX 77339–3097

Proper correction for axial dispersion in gel permeation chromatography (GPC, or SEC for size exclusion chromatography) has not been adequately addressed in the available commercial softwares today. In this paper, we will propose a new computer algorithm to address this problem. Our approach is built on the theoretical foundations previously described in the GPCV2 paper published by Yau, et. al (1), where GPCV2 stands for an improved version, or a 2nd version of the well-known Hamielec's broad-standard GPC calibration method (2). Most commercial GPC softwares provide the dispersion-correction capability of the Hamielec method. Very few of them have adopted the GPCV2 improvement. While the GPCV2 method is better, it still has its problems. GPCV2 corrects for the axial dispersion effect in the calculated molecular weight values of polymer samples, but it does not remove the axial dispersion effect in the reported molecular weight distribution (MWD) curve of the sample. Our computer algorithm at Chevron removes this discrepancy. In this paper, we will present the theoretical basis for two Chevron methods of GPC calibration and data processing.

With the 5-μm and 10-μm high-resolution GPC packings that are widely available nowadays, there has been a common tendency among GPC practitioners to disregard the effect of column axial dispersion or instrumental band-broadening problems in conventional GPC experiments. In many applications, this neglect of axial dispersion correction is quite justifiable. The reasons for the present author to revisit the subject of axial-dispersion correction are the following: (1) For the polyolefin work in the author's laboratory, samples with the high-MW-end of the MWD curves extending up to 10 million MW are not uncommon. In this case, in order to avoid shear degradation, there is little choice but to use larger particle size (20-μm) columns and accept the low resolution condition of the GPC experiment. (2) There is a need for GPC software that corrects the sample MWD curve for the axial-dispersion effect. In industry, the overlay of MWD curves often is more useful for solving polymer problems than the tabulation of average-

MW values. But, without a band-broadening correction, the MWD-overlay approach is affected by the column resolution changes that occur when columns deteriorate with time, or whenever an old column set is replaced. (3) Broad-standard calibration, when used improperly, can lead to unacceptably large MW errors even when high-resolution columns are used. Unexpectedly large errors in the MWD curves can also occur.

Figure 1 is a quick overview of the choice of GPC calibration algorithms commonly available in the commercial software today. These algorithms are briefly discussed below. In the discussions that follow, we use the word "standard" to represent the calibration standard of know molecular weight, and the word "sample" to represent a polymer sample with the molecular weight of which is being determined.

If several narrow standards of known MW of the same chemical structure as the sample are available, the true calibration line for a GPC system [i.e., $M_T(v)$ with the true calibration constants of D_1 and D_2] can be obtained experimentally by using the GPC peak positions of the narrow standards. The sample MW values calculated using this narrow-standard calibration are not corrected for the axial dispersion effect. But, these MW errors are often very small for high-resolution columns. These errors are also highly predictable and symmetrical, i.e., the Mw value will be slightly too high, and the Mn value will be slightly too low by roughly the same percentage. Although this method does not correct the sample MWD curve for axial dispersion, it would not cause gross errors of shifting the MWD curves to wrong MW regions, as it could in the case of the Hamielec method described below. The problem of the narrow-standard method is its limited applicability to just a handful of polymer types where there are narrow standards commercially available.

The Hamielec broad-standard method uses the known Mw and Mn values of a broad standard and has the computer search for an effective calibration line [i.e., $M_H(v)$ with the effective calibration constants of D_1' and D_2'] from the experimental GPC elution curve $F(v)$ of the broad standard. Since the experimental elution curve of the standard has been affected by axial dispersion, the Hamielec method partially compensates for the axial dispersion error if the sample is of the similar MW and MWD features as the standard. However, this method can lead to gross inaccuracy of sample MW values if the elution volume of the sample deviates substantially from that of the broad standard. The axial dispersion in the broad standard causes the Hamielec $M_H(v)$ to rotate counter-clockwise from the true calibration curve $M_T(v)$. The pivot point is near the center of the broad-standard elution curve. To the left of this point, $M_H(v)$ is lower than $M_T(v)$ and therefore will underestimate the sample MW. Conversely, for samples that elute later than the broad standard, the Hamielec method can grossly overestimate the sample MW. For the same reasons, this rotation of $M_H(v)$ can also cause erroneous shifts of the sample MWD curves to the wrong MW regions. Therefore, the MW errors caused by axial dispersion becomes a much more complex problem and a much less predictable challenge in the Hamielec method as compared to the narrow-standards method. Because of this reason, the axial dispersion problem of the Hamielec method is highly dependent on the polydispersity or the Pd value of the broad standard [Pd is defined as the ratio of M_w/M_n]. The broader the standard the better, with less rotation of the $M_H(v)$ line. For polyolefins, the use of broad standards having polydispersity value of 10 or more is

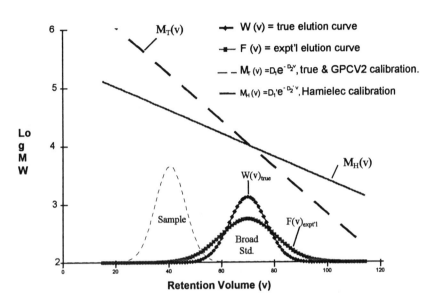

Calibration Methods	M_w, M_N, Pd Corrections	MWD-overlay Precision	Remarks
• Narrow Standards - peak position calibration curve	No	Poor	Limited to polymer samples of same structures as the std.
• Broad Standard -Hamielec method $M_w=D_1 \Sigma F(v)e^{-D_2 \cdot V}$ $M_N=D_1 / \Sigma F(v)e^{+D_2 \cdot V}$	Partial	Very Poor	Correction limited to samples of similar MWD as the standard
• Broad Standard - duPont GPCV2 $M_w=D_1 e^{-(D_2\sigma)^2/2} \Sigma F(v)e^{-D_2 V}$ $M_N=D_1 e^{+(D_2\sigma)^2/2} \Sigma F(v)e^{+D_2 V}$	Yes	Poor	Corrections apply to samples of any MWD

where , $F(V)_{exp\,t'l} = \int W(V)_{true} \cdot G(V - y) \cdot dy$

and, σ = the standard deviation of a Gaussian $G(V-y)$ axial dispersion function.

Figure 1. Axial Dispersion Problem in Existing GPC Softwares.

possible. The situation is more difficult for analyzing condensation polymers like Nylon, Dacron, and so on, where the polymer polydispersity values are quite low, staying around 2 or so. The errors of the Hamielec method using a broad standard of polydispersity of 2 can be extremely high and risky, unless all samples are of similar MW and MWD to that of the standard. For such cases, one should consider creating a broad standard of much higher polydispersity by blending together several standards of very different MW values.

The GPCV2 method uses mathematical formulations (see Figure 1) of Mw and Mn that have already accounted for the column axial dispersion effect and therefore allows the search for the true calibration curve $M_T(v)$ and the true calibration constants of D_1 and D_2 parameters, where the effect of axial dispersion is measured by the standard deviation σ of a narrow standard GPC elution peak. This GPCV2 method represents an attempt to minimize the counter-clockwise rotation of the broad-standard calibration curve, and thus to minimize the MW accuracy problems created by the calibration line rotation. The method provides the correction of the sample Mw, Mn and Pd values, but still provides no correction to the MWD curve. A glaring discrepancy exists in GPCV2 where the calculated sample MW values are now closer to the true values, but the calculated MWD display remains too broad as compared to the true MWD of the sample. This discrepancy is removed in the Chevron methods reported below.

CHEVRON METHOD-1

Chevron Method-1 is a direct extension of the GPCV2 method by introducing one additional step of an "effective sample calibration search" for the M(v) line of every sample with its own calibration constants of the D_1" and D_2" parameters. The method is outlined in Figure 2.

The thought behind Chevron-1 is as follows, referring to Figure 2. One can expect that, much like the behavior of the broad standard in the Hamielec method, every polymer sample would prefer to have its own effective MW-calibration curve that would be consistent with the sample's true MW values under the existing axial dispersion conditions. Therefore, for every sample, there will be a unique M(v) curve that will have its rotation and pivot point located just right to produce the correct Mw and Mn values for the sample. The effect of applying this M(v) calibration line [which is less steep than the true calibration line $M_T(v)$] to the experimental F(v) elution curve of the sample would then act as if it is compressing the peak width of the F(v) curve into a narrower MW window, and therefore result in a sharper and narrower sample MWD curve. The result is that the axial dispersion effect on the sample F(v) curve is now removed from the calculated sample MWD curve. This correction of axial dispersion in sample MWD helps to make it more meaningful to compare sample MWD-overlay curves from GPC results obtained under different column resolution conditions.

The first 3 steps in the Chevron-1 algorithm outlined in Figure 2 are the same steps that exist in GPCV2. Steps 4 and 5 are added here to achieve the desired effect of removing axial dispersion from the sample MWD curve.

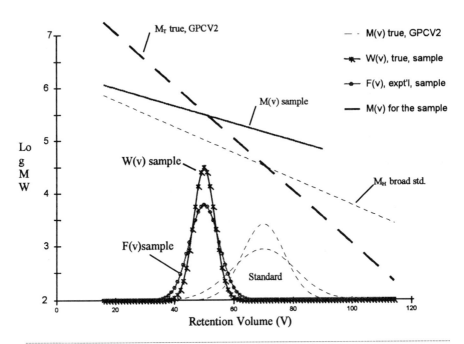

Calibration Method	Mw, Mn, Pd Corrections	MWD-overlay Precision	Remarks
Chevron-1	Yes	Better	Corrections apply to bell-shaped MWD and linear columns

Chevron-1 Algorithms:
1. Use narrow-standard calibration if available, or, obtain GPCV2 calibration line from broad standard.
2. Calculate $M_{W, expt'l}$ and $M_{N, expt'l}$ from $F(V)_{sample}$.
3. Calculate $M_{W, true}$ and $M_{N, true}$ of the sample from:
   $$M_{W, true} = e^{-(D_2\sigma)^2/2} M_{W, expt'l}, \text{ and, } M_{N, true} = e^{+(D_2\sigma)^2/2} M_{N, expt'l}.$$
4. Search the sample M(v) function [every sample has its own unique $M(v) = D_1 e^{-D_2"v}$]:
   $$M_W = D_1" \Sigma F(v) e^{-D_2"v}, \text{ and, } M_N = D_1" / \Sigma F(v) e^{+D_2"v}.$$
5. Calculate sample MWD curve from F(v) and M(v) of the sample.

Figure 2. Chevron Method-1 for GPC Axial Dispersion Correction.

One may notice that the assumption of linear calibration is implicit in the Hamielec method, GPCV2, and also in Chevron-1. The linear calibration assumption is made to improve the precision of these methods, although, at the expense of the versatility of the methods. For instance, the Chevron-1 method described here is expected to work well with the usual bell-shaped GPC elution curves and linear columns. But, for samples with elution curves very different from the usual bell-shaped GPC curve, one may need to consider using the Chevron-2 method described below. The Chevron-2 method is more easily applicable to situations of non-linear calibration curve as well. However, it is all a matter of compromise: Chevron-2 is more versatile, but Chevron-1 gives better precision and is less affected by detector noises.

CHEVRON METHOD-2

The linear calibration approximation used in Chevron-1 is not expected to work well, for example, with a bimodal MWD sample as illustrated in Figure 3. Each component in this bimodal elution curve would prefer to have its own "effective" calibration line. Since the pivot points of the two components are separated, there will be a discontinuous break of the M(v) curve in-between the two component peaks. Therefore, the shape of the overall "effective" MW-calibration curve can be quite complicated. Fortunately, the formulations developed in the GPCV2 paper (1) can be used to estimate the shape of these "effective" MW-calibration curves for samples of any MWD and GPC elution curve shapes. These formulations are shown in Figure 3, step 2 of the Chevron-2 algorithm, where Mw(v) and Mn(v) stand for the local weight-average and number-average MW values at any retention volume v along the sample GPC elution curve. And, F(v) stands for the experimental GPC elution curve at the retention volume v, while $F(v-D_2\sigma^2)$ stands for the experimental GPC elution curve at a retention volume of $(v-D_2\sigma^2)$.

Same as the Chevron-1 method, Chevron-2 also has the effect of removing axial dispersion from MWD by compressing the F(v) curve into a narrower MW region by way of the counter-clockwise rotation of the "effective" MW-calibration curve of the sample. However, in addition to narrowing down the overall MWD breadth of the MWD window, Chevron-2 also improves the resolution of the MWD curve. This resolution enhancement effect can be explained by a close examination of the bimodal example in Figure 3. The M(v) curve of this bimodal sample can be approximated by the $M_G(v)$ curve calculated in step 3 of Chevron-2 algorithm. The geometric-average $M_G(v)$ curve is situated in-between the Mw(v) and Mn(v) curves shown in the graph. One sees that, this $M_G(v)$ curve is flat on the two sides, but has a steep slope in the transition region between the two pivot points. We know from the previous explanation of the Chevron-1 method that the flat slope at the two ends would have the effect of causing the narrowing of the overall MWD curve. The opposite is true in the middle overlapping part of the bimodal peaks where the $M_G(v)$ slope is steep. In this middle portion, the spacing between the F(v) data points will be more stretched out to cause a valley to be developed in the middle of the sample MWD curve. All these manipulations are automatically taken care of by digital

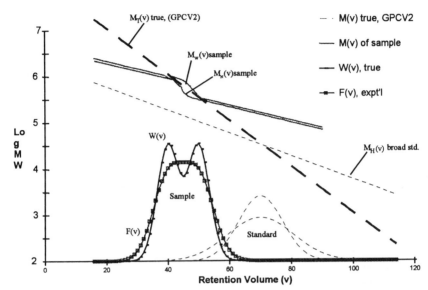

Calibration Method	M_W, M_N, Pd Corrections	MWD-overlay Precision	Remarks
Chevron-2	Yes	Best?	Corrections apply to samples of any MWD, possible with non-linear columns, has resolution-enhancement feature, but more sensitive to baseline noise than Chevron-1.

Chevron-2 Algorithms:
1. Use narrow-standard calibration(linear or non-linear) if available, or, obtain GPCV2 calibration line from broad standard.
2. For every sample , calculate:

$$M_W(v) = \frac{F(v - D_2\sigma^2)}{F(v)} \cdot e^{+\frac{1}{2}(D_2\sigma)^2} \cdot M_T(v) \qquad M_N(v) = \frac{F(v)}{F(v + D_2\sigma^2)} \cdot e^{-\frac{1}{2}(D_2\sigma)^2} \cdot M_T(v)$$

$$\overline{M}_{W,sample} = \sum F(v)M_W(v) \qquad \overline{M}_{N,sample} = 1 \div \left\{ \sum \left[F(v)/M_N(v) \right] \right\}$$

3. Calculate the geometric average molecular weight:
$$M_G(v) = \left(\sqrt{M_W(v) \cdot M_N(v)} \right)$$
4. Calculate sample MWD curve from F(v) and $M_G(v)$ of the sample.

Figure 3.   Chevron Method-2 for GPC Axial Dispersion Correction.

computation following the step 4 of the algorithm using the equation (1) described below. Since we are no longer dealing with linear approximation, the local slope of the $M_G(v)$ curve will have to be accounted for in step 4 to calculate the sample MWD curve. The desired y-axis value of a MWD curve would have to be calculated rigorously using the following equation:

$$\frac{dW}{d\log M} = \frac{dW}{dV} \bullet \frac{dV}{d\log M} = \frac{dW}{dV} \Bigg/ \frac{d\log M}{dV} = \frac{F(v)}{S(v)} \tag{1}$$

where $S(v)$ is the slope of the log $M_G(v)$ calibration curve at the retention volume v. Since $S(v)$ is sensitive to the signal-to-noise level of the GPC detector, the log $M_G(v)$ calibration curve needs be curve-fitted segmentally to improve the precision of the $S(v)$ calculation under noisy detector conditions.

DISCUSSION

In summary, the key issues this manuscript tries to address are the following: (1) High-precision GPC-MWD overlay curves are more useful than the tabulated Mw, Mn, and polydispersity values for solving industrial polymer problems. (2) GPC column efficiency deteriorates, as the column axial dispersion effect becomes worse with time. Currently, it is very difficult to obtain good MWD overlay curves for samples that are GPC-analyzed on different days, in different weeks, or in different months. (3) Chevron-1 or -2 has the axial dispersion correction feature for the sample MWD curves to allow more meaningful comparison of MWD-overlay curves obtained at different times.

Although Chevron-2 is more accurate than Chevron-1 and also adds some "resolution-enhancement" or "deconvolution effect" to the final MWD, it has its drawback of being more sensitive to detector noise than Chevron-1. Therefore, the choice of which method to use depends largely on the nature of the sample elution curve shape and the GPC detector signal-to-noise conditions. In theory, Chevron-2 is also applicable for nonlinear calibration curve by using the local calibration slope of $D_2(v)$ at every retention volume. Such practice however will add more detector noise effects and more precision problems to the calculations when dealing with real experimental data.

What are presented in this paper are two theoretical models of GPC axial dispersion correction. The current models were developed based on a symmetrical Gaussian axial dispersion function of a constant σ-value for the axial dispersion standard deviation. The application of the methods to real data to study the effect of detector noise on the two methods will be included in a future report. Further complications may occur in real data applications, the σ-value can change with elution volume, especially at high molecular weights (and with low resolution columns!). Also, the axial dispersion function may sometimes deviate from Gaussian to have a skewed component. These problems are real and they are very important issues that need be addressed for future improvements over the two current methods described in this paper.

REFERENCES

1. Yau, W. W., Stoklosa, H. J., and Bly, D. D., J. Polym. Sci., **1977,** 21, p.1911.

2. Balke, S. T., A. E. Hamielec, A. E., LeClair, B. P., and Pearce, S. L., Ind. Eng. Chem., Prod. Res. Develop. **1969,** 8, p.54.

Chapter 4

Axial Dispersion Correction for the Goldwasser Method of Absolute Polymer M_n Determination Using SEC-Viscometry

Wallace W. Yau

Chevron Chemical Company LLC, Kingwood Technical Center, 1862 Kingwood Drive, Kingwood, TX 77339–3097

Copolymers and polymer blends, especially where there is compositional drift across the sample molecular weight distribution (MWD), present serious problem to the absolute MW determination by conventional size exclusion chromatography (SEC or GPC for gel permeation chromatography) using only a single concentration detector, which is generally affected by the sample chemical compositional differences. Since sample specific refractive index increment (dn/dc) changes with chemical composition, MW determination of these kinds of samples also presents difficulty for SEC using an on-line light scattering detector. A complete MWD of polymer samples with compositional variances is difficult. However, the introduction of a SEC-viscometry method, known as the Goldwasser Method (1) makes it possible to determine a number-average molecular weight value (Mn) of polymer samples with complex compositional heterogeneity. The Goldwasser method is based on the SEC universal calibration principle. This unique SEC method uses only the SEC data collected on an on-line viscosity detector to determine the Mn of polymer, regardless of its chemical structural differences. The method requires no concentration detector. In this work, an axial dispersion correction factor for the Goldwasser Mn value is developed. The theoretical basis of this correction factor is derived from the use of a Gaussian instrument band-broadening function as the axial dispersion peak shape model. Substantial improvement in accuracy of the Goldwasser Mn value is expected with this correction factor. It is hoped that this improvement will help to make this important method more widely useful in the polymer characterization field.

BACKGROUND

(1) Universal Calibration (UC) Principle (assuming no axial dispersion)

At any GPC elution volume (e.g. the j-th retention volume), all the i-th polymer molecules will co-elute at this j-th retention volume if,

$$M_1[\eta]_1 = M_2[\eta]_2 = \ldots = M_i[\eta]_i = H_j \tag{1}$$

where M, $[\eta]$, and H, are the polymer molecular weight (MW), intrinsic viscosity (IV), and hydrodynamic volume (HV), respectively. Or,

$$\text{Or,} \quad [\eta]_1 = \frac{H_j}{M_1}, \quad [\eta]_2 = \frac{H_j}{M_2}, \quad \ldots, \quad [\eta]_i = \frac{H_j}{M_i} \tag{2}$$

The observed intrinsic viscosity at the j-th GPC retention volume is the weighted-average of the intrinsic viscosity values of all the i-th molecules co-eluting at that retention volume:

$$[\eta]_{wj} = \frac{\sum_i C_i[\eta]_i}{\sum_i C_i} = \frac{\sum_i \dfrac{C_i H_i}{M_i}}{\sum_i C_i} = \frac{H_j}{\sum_i C_i / \sum_i (C_i / M_i)} = \frac{H_j}{M_{nj}} \tag{3}$$

$$\text{Or,} \quad (M_n \cdot [\eta]_w)_j = H_j \tag{4}$$

$$\text{where} \quad H_1 = H_2 = \ldots = H_i = H_j \tag{5}$$

What is important in equation 3 is that the proper molecular weight average to use with universal calibration methodology is the number-average molecular weight. Equation 4 is an important theoretical result developed by Hamielec and Ouano (2). Equation 4 and 5 shows the proper statistical averages of MW and IV that are required to treat the universal calibration result of heterogeneous polymer samples at every GPC retention volume, assuming no GPC axial dispersion. With axial dispersion, the assumption stated in Equation 5 can no longer to be true (see equation 8 in the later theoretical results section).

Current Goldwasser Method (1)

$$\overline{M}_{n,Bulk\,Sample} = \frac{\sum_j C_j}{\sum_j \left(\dfrac{C_j}{M_{nj}}\right)} = \frac{\sum_j C_j}{\sum_j \left(\dfrac{C_j}{H_j / [\eta]_{wj}}\right)}$$

$$= \frac{\sum_j C_j}{\sum_j \left(\dfrac{C_j[\eta]_{wj}}{H_j}\right)} = \frac{\sum_j C_j}{\sum_j \left(\dfrac{\eta_{sp}}{H}\right)_j} \tag{6}$$

where, η_{sp} is the sample specific viscosity.

Equation 6 gives the Mn calculation in the conventional SEC-viscometry method with universal calibration, i.e., where the intrinsic viscosity is measured at each elution slice. The numerator in equation 6 is proportional to the sample weight injected and the ηsp value in the denominator is proportional to the excess pressure-drop of an online viscometer signal. Based on these observations, Goldwasser (1) derived equation 7 below to show that the bulk sample Mn value can be obtained by using only the viscosity detector signal, without the use of the concentration detector signal.

$$\overline{M}_{n,Bulk Sample} = \frac{Sample Weight Injected}{\sum_j (\frac{Viscosity Detector Signal}{H})_j} = \overline{M}_{n,Goldwasser} \tag{7}$$

Under the condition of accurate sample weight injected and total sample recovery, a same Mn value is obtained either by the conventional SEC-viscometry of equation 6, or by the Goldwasser method of equation 7.

Comments:
(1) Positive Features: this Mn determination is generally applicable to all polymer structural heterogeneity, that includes polymer branching, chain rigidity, polymer blends and copolymers with compositional and chemical structural differences, as long as that the GPC separation behaves normally and obeys the universal calibration principle.
(2) The current Goldwasser method does not account for the reality of the GPC axial dispersion problem. The problem is that, due to mixing effects, not all molecules eluted at the same retention volume have exactly the same hydrodynamic volume. Equation 5 is valid only if there is no axial dispersion and non-size-exclusion effects.

THEORETICAL RESULTS

Under axial dispersion circumstances, the hydrodynamic volume values of different molecules co-elute at the same SEC retention volume are no longer all of the same value. With axial dispersion, there will be local polydispersity in the hydrodynamic volume values within each individual SEC retention volume. The challenge is to question what hydrodynamic volume average should be used to account for the axial dispersion effects. The presence of axial dispersion affects the hydrodynamic volume. In order to understand how axial dispersion affects the Mn value calculated from the SEC-viscometry data, it is important to understand first how axial dispersion affects the relationship of SEC hydrodynamic volume versus the retention volume. The mathematical derivations of these concepts are presented below in the four inter-connected sections.

(1) Generalization of UC Equation to Include Axial Dispersion and Non-SEC Perturbations

$$[\eta]_{wj} = \frac{\sum_i C_i [\eta]_i}{\sum_i C_i} = \frac{\sum_i (C_i \frac{H_i}{M_i})}{\sum_i C_i}$$

$$= \frac{\sum_i (\frac{C_i}{M_i} H_i)}{\sum_i (\frac{C_i}{M_i})} \cdot \frac{\sum_i (\frac{C_i}{M_i})}{\sum_i C_i} = \left(\frac{H_n}{M_n}\right)_j \tag{8}$$

Or, $\quad (M_n \cdot [\eta]_w)_j = (H_n)_j \tag{9}$

where H_n which equals to the first part of equation 8 is defined as the number-average hydrodynamic volume of the polymer molecule. Unlike equation 4, equation 9 is more generally applicable for all cases where not all the i-th molecules eluting at the j-th retention volume have the same hydrodynamic volume, i.e., in equation 8, not all $H_i = H_j$.

(2) Experimental Evaluation of $H_n(v)$

$$(H_n)_j = \left(\frac{\sum\left(\frac{C_i}{M_i} \cdot H_i\right)}{\sum_i\left(\frac{C_i}{M_i}\right)}\right)_j = \left(\frac{\sum_i(C_i[\eta]_i)}{\sum_i\left(\frac{C_i[\eta]_i}{H_i}\right)}\right)_j = \left(\frac{\sum_i(\eta_{sp})_i}{\sum_i\left(\frac{\eta_{sp}}{H}\right)_i}\right) \tag{10}$$

Equation (10) provides a means to experimentally evaluate $H_n(v)$ as a function of retention volume v, through the use of the viscosity detector signal <u>alone</u>, without the use of the RI or other concentration detector signal.

(3) Effect of Gaussian Axial Dispersion on $M_n(v)$

We let $W(v)$ and $W_\eta(v)$ to represent the true concentration and viscosity detector elution profile respectively, that has not been affected by axial dispersion. And, we let $F(v)$ and $F_\eta(v)$ to represent the experimental concentration and viscosity detector elution curve respectively, that has been affected by axial dispersion. And, we let G(v-y) to represent a Gaussian axial dispersion function having a standard deviation of σ for the axial dispersion sigma value, expressed in retention volume units. The effect of axial dispersion on the local Mn values at every retention volumes can be formulated as the following.

$M_n(v) = the \ "M_n value" of \ polymer \ observed \ at \ the \ retention volume" \ v"$

$$= \frac{\int_{-\infty}^{+\infty} W(y) \cdot G(v-y)dy}{\int_{-\infty}^{+\infty} W(y) \cdot G(v-y)/M_n(y)dy} = \frac{\int_{-\infty}^{+\infty} W(y) \cdot G(v-y)dy}{\int_{-\infty}^{+\infty} \dfrac{W(y) \cdot G(v-y)}{H_t(y)} \cdot [\eta](y)dy}$$

$$= \frac{\int_{-\infty}^{+\infty} W(y) \cdot G(v-y)dy}{\int_{-\infty}^{+\infty} \dfrac{[W(y) \cdot [\eta](y)] \cdot G(v-y)}{H_t(y)} dy}$$

$$= \frac{\int_{-\infty}^{+\infty} W(y) \cdot G(v-y)dy}{\int_{-\infty}^{+\infty} [W(y) \cdot [\eta](y)] \cdot G(v-y)dy} \cdot \frac{\int_{-\infty}^{+\infty} [W(y) \cdot [\eta](y)] \cdot G(v-y)dy}{\int_{-\infty}^{+\infty} \dfrac{[W(y) \cdot [\eta](y)] \cdot G(v-y)}{H_t(y)} dy}$$

$$= \frac{1}{[\eta]_w(v)} \cdot \left[\frac{\int_{-\infty}^{+\infty} W_\eta(y) \cdot G(v-y)dy}{\int_{-\infty}^{+\infty} \dfrac{W_\eta(y) \cdot G(v-y)}{H_t(y)} dy} \right] \equiv \frac{H_{n\eta}(v)}{[\eta]_w(v)} \tag{11}$$

The quantity $H_{n\eta}(v)$ is defined as the argument in the square brackets in equation 11. By applying the mathematical analogy of equation A-9 given in a previous paper (3), we have:

$$H_{n\eta}(v) = \frac{F_\eta(v)}{F_\eta(v+H_2\sigma^2)} \cdot e^{-\frac{1}{2}(H_2\sigma)^2} \cdot H_t(v) \tag{12}$$

Therefore,

$$M_n(v) = \frac{H_{n\eta}(v)}{[\eta]_w(v)} = \frac{1}{[\eta]_w(v)} \cdot \left(\frac{F_\eta(v)}{F_\eta(v+H_2\sigma^2)} e^{-\frac{1}{2}(H_2\sigma)^2} \right) \cdot H_t(v) \tag{13}$$

where $H_t(v)$ represents the true linear UC calibration line not affected by axial dispersion,

$$H_t(v) = H_1 \cdot e^{-H_2 v} \tag{14}$$

(4) Effect of Gaussian Axial Dispersion on The $\overline{M}n$ Value of The Bulk Sample

The substitution of the Mn(v) expression of equation 13 into the definition of the bulk sample Mn calculation, we have:

$$\overline{M}_{n,\,Sample} = \frac{\int_{-\infty}^{+\infty} F(v)\,dv}{\int_{-\infty}^{+\infty} \dfrac{F(v)}{M_n(v)}\,dv} = \frac{\int_{-\infty}^{+\infty} F(v)\,dv}{\int_{-\infty}^{+\infty} \dfrac{F(v)}{\dfrac{1}{[\eta](v)} \cdot \dfrac{F_\eta(v)}{F_\eta(v+H_2\sigma^2)} \cdot e^{-\frac{1}{2}(H_2\sigma)^2} \cdot H_t(v)}\,dv}$$

$$= e^{-\frac{1}{2}(H_2\sigma)^2} \cdot \frac{\int_{-\infty}^{+\infty} F(v)\,dv}{\int_{-\infty}^{+\infty} \dfrac{F(v)\cdot[\eta](v)}{H_t(v)}\left(\dfrac{F_\eta(v+H_2\sigma^2)}{F_\eta(v)}\right)dv}$$

$$= e^{-\frac{1}{2}(H_2\sigma)^2} \cdot \frac{\int_{-\infty}^{+\infty} F(v)\,dv}{\int_{-\infty}^{+\infty} \dfrac{F_\eta(v)}{H_t(v)}\left(\dfrac{F_\eta(v+H_2\sigma^2)}{F_\eta(v)}\right)dv}$$

$$= e^{-\frac{1}{2}(H_2\sigma)^2} \cdot \frac{\int_{-\infty}^{+\infty} F(v)\,dv}{\int_{-\infty}^{+\infty} F_\eta(v+H_2\sigma^2)\cdot\left(\dfrac{1}{H_1}\right)\cdot e^{H_2 v}\,dv}$$

$$\overline{M}_{n,\,sample} = e^{-\frac{1}{2}(H_2\sigma)^2} \cdot \frac{\int_{-\infty}^{+\infty} F(v)\,dv}{\int_{-\infty}^{+\infty} F_\eta(v)\cdot\left(\dfrac{1}{H_1}\right)\cdot e^{H_2(v-H_2\sigma^2)}\,dv}$$

$$= e^{-\frac{1}{2}(H_2\sigma)^2} \cdot \frac{\int_{-\infty}^{+\infty} F(v)\,dv}{\int_{-\infty}^{+\infty} F_\eta(v)\cdot\left(\dfrac{1}{H_1\cdot e^{-H_2 v}}\right)dv} \cdot e^{+(H_2\sigma)^2}$$

$$= e^{+\frac{1}{2}(H_2\sigma)^2} \cdot \left(\frac{\int_{-\infty}^{+\infty} F(v)\,dv}{\int_{-\infty}^{+\infty} \dfrac{F_\eta(v)}{H_t(v)}\,dv}\right) = e^{+\frac{1}{2}(H_2\sigma)^2} \cdot \overline{M}_{n,\,Goldwasser} \qquad (15)$$

It is noted that the expression in the square bracket in equation 15 is the same expression (except that it is written in the integral notation here) for the Mn value of the conventional SEC-viscometry, and the Mn value of the Goldwasser method described previously in equation 6 and 7.

Or,
$$\overline{M}_{n,True} = e^{+\frac{1}{2}(H_2\sigma)^2} \cdot \overline{M}_{n,Goldwasser} \tag{16}$$

with, the proper correction factor = $e^{+\frac{1}{2}(H_2\sigma)^2}$. \hfill (17)

Thus, $e^{+\frac{1}{2}(H_2\sigma)^2}$ is the proper factor to correct the apparent or the Goldwasser number-average molecular weight to an axial-dispersion corrected or true number-average molecular weight obtainable from SEC-viscometry and universal calibration methodology.

DISCUSSION

The result of this work (equation 16) indicates that, without axial dispersion correction, the current Goldwasser method of sample Mn determination using GPC-viscometry would underestimate the true Mn value for the sample. This same correction factor is also applicable to the Mn value calculated directly in the conventional SEC-viscometry approach using universal calibration where intrinsic viscosity is measured at each elution slice.

It is interesting that the Mn correction factor $e^{+\frac{1}{2}(H_2\sigma)^2}$ is not a function of the sample elution curve shape. That means, the percentage error caused by symmetrical axial dispersion as approximated here by a Gaussian model will be the same for all samples, regardless of their MWD curve shape or their chemical or structural heterogeneity.

The correction factor here is developed on the model of symmetrical Gaussian axial dispersion function and linear universal calibration curve. This correction factor is expected to work well in GPC experiments using linear columns with good column efficiency. It is possible to generalize the correction factor for non-linear UC calibration and skewed axial dispersion function if the need exists.

To obtain proper correction of the Mn value, it is important that the viscosity detector volume-offset should have been properly aligned with the concentration detector signal in the software, and thus, aligned with the position of the universal calibration curve. Improper volume-offset alignment in the software can affect the Goldwasser Mn values. In fact, a small intentional shift of the viscosity GPC tracing to a shorter retention time will tend to adjust the Goldwasser Mn values upward, closer to the true Mn value. It has been shown that an intentional shift in viscosity detector signal to shorter retention is equivalent to the effect of correcting for symmetrical axial dispersion in GPC-viscometry (4).

ACKNOWLEDGMENT

The author is deeply indebted to Professor S. T. Balke and Dr. T. H. Mourey who have kindly pointed out that they have also obtained the same Goldwasser Mn correction factor by using a very different mathematical approach (5). It is the hope of this author that readers may still find the insights behind the mathematical derivations presented in this manuscript somewhat interesting and contributing to the better understanding of the problem.

REFERENCES

1. Goldwasser, J. M., in *Chromatography of Polymers: Characterization by SEC and FFF;* Provder, T., Ed.; ACS Symposium Series 521; American Chemical Society: Washington, D.C.; 1993, p.243.
2. Hamielec, A. E., and Ouano, A. C., J. Liq. Chromatogr., **1978,** 1, p.111.
3. Yau, W. W., Stoklosa, H. J., and Bly, D. D., J. Appl. Polym. Sci., **1977,** 21, p.1911, equations A7-A9.
4. Yau, W. W., in *Proceedings of the International GPC Symposium* '87, published by Waters Corporation, Milford, MA, **1987,** p. 148.
5. Balke, S. T., Mourey, T. H., and Harrison, C. A., J. Appl. Polym. Sci., **1994,** Vol. 51, pp. 2087-2102.

Chapter 5

Polymer Characterization by High Temperature Size Exclusion Chromatography Employing Molecular Weight Sensitive Detectors

S. J. O'Donohue and E. Meehan

Polymer Laboratories Ltd., Essex Road, Church Stretton, Shropshire SY6 6AX, United Kingdom

The use of molecular weight sensitive detectors, in particular light scattering and viscometry, has become commonplace in the application of size exclusion chromatography (SEC) to the characterization of polymers. Both detectors can facilitate accurate determination of molecular weight distribution when coupled to SEC. High temperature SEC is used to analyse a range of engineering polymers whose properties are such that dissolution can only be achieved in aggressive solvents at temperatures in excess of 135°C. SEC equipment for this type of analysis is generally very specialised and demands a number of design features which are necessary to achieve acceptable chromatography. The incorporation of molecular weight sensitive detectors into such equipment is a vital tool in molecular characterization experiments. Typical applications include polyolefin polymers whose analysis is illustrated in trichlorobenzene at temperatures in the range 135-180°C. A more demanding application, the characterization of poly(phenylene sulfide), in o-chloronaphthalene at 210°C is studied in detail.

Instrumentation for high temperature SEC experiments is generally purpose built to encompass some basic design requirements including a constant flow rate solvent delivery system, a temperature controlled oven compartment in which the SEC columns and detection systems can be housed and an automated high temperature sample injection system.

The use of on-line molecular weight sensitive detectors, light scattering and viscometry, has become commonplace in SEC because not only do they facilitate the determination of molecular weight distribution, they can also provide information on polymer conformation e.g. branching. Ideally these detectors also need to be housed

© 1999 American Chemical Society

in the main SEC instrument in order to avoid the spurious effects of temperature variation throughout the system which can occur when individually temperature controlled modules are employed (1). Figure 1 illustrates the arrangement of the three detectors in a modern integrated high temperature SEC system (PL-GPC210, Polymer Laboratories, UK). The injection system and oven compartment can be temperature controlled up to 220°C to accommodate demanding applications.

Commercial instrumentation for high temperature SEC has until recently been limited in temperature capability to around 145°C. Consequently very little work has been published on polymer characterization by SEC at temperatures in excess of 145°C, despite the fact that the increase in temperature may offer significant benefits in the analysis of very high melting point polyolefins. However analysis of polymers at temperatures in excess of 145°C has been reported using a home built system for the characterization of poly(phenylene sulfide), although severe restrictions in the instrument design could not be overcome and compromises in detector choice and injection of the samples had to be made (2). This paper describes the use of a modern, commercial high temperature SEC instrument applied to polyolefin analysis at higher temperatures employing both viscometry and light scattering detection. A case study describing the characterization of poly(phenylene sulfide) at 210°C is discussed in detail. Poly(phenylene sulphide) (PPS) is a semi-crystalline polymer possessing a combination of properties desirable to a designer making it an important engineering thermoplastic. There are no known solvents for PPS below 200°C and pioneering work has been published (2) describing the use of 1-chloronaphthalene as a suitable solvent for the determination of solution properties at 208°C. The characterisation of PPS samples by SEC-viscometry is presented here.

Experimental

The PL-GPC210 high temperature SEC system contains a differential refractive index (DRI) detector as standard. The instrument used extensively in this study was also fitted with a four capillary bridge viscometer (Model 210R, Viscotek, USA). A second system in which a viscometer and a light scattering detector (model PD2040, Precision Detectors, USA) were installed was also used.

In both systems the viscometer was connected in parallel with the DRI detector to give a flow split between the two detectors of 55:45 (viscometer:DRI). In the triple detector system the light scattering detector was connected in series before the other two detectors as illustrated in Figure 2.

SEC separations were performed using a column bank of three PLgel 10µm MIXED-B 300 x 7.5 mm columns and the eluent flow rate was maintained at 1 ml/min. A flushed full loop injection of 200µl was employed throughout the study. The eluents and temperatures studied are summarised in Table I. All solvents were analytical reagents and were used without any further purification.

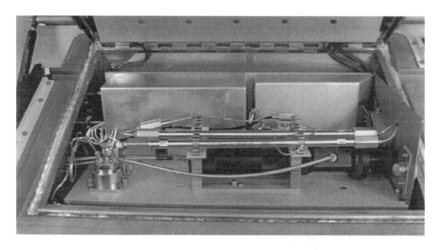

Figure 1.   Triple Detector Arrangement in a PL-GPC210

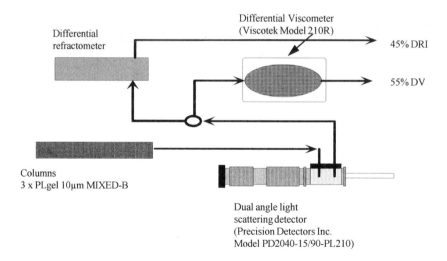

Figure 2.   Detector Configuration

Table I. SEC Eluents and Temperatures of Operation

Solvent	Temperatures (°C)
Tetrahydrofuran (THF)	40
1,2,4-trichlorobenzene (TCB)	135, 145, 150, 160, 180
o-chloronaphthalene (CN)	180, 210

Stabilised THF was used as received (250ppm BHT) and BHT was added to the TCB at a concentration of 0.00125%.

Narrow polydispersity polystyrene standards (Polymer Laboratories, UK) were used for column and instrument calibration. Solutions were prepared and allowed to dissolve at ambient temperature overnight before use. Polyethylene standards (NIST, USA) were dissolved in TCB and CN at 160°C for 1-2 hours. The commercial poly(phenylene sulphide) sample solutions were prepared in CN at 230°C for 1 hour. In all cases the polymer concentration was known accurately and was of the order of 0.5-2.0 mg/ml depending on molecular weight.

All data manipulation was performed using PL Caliber SEC software (Polymer Laboratories, UK). The volumetric offset or interdetector delay (IDD) between the various detectors was calculated from the retention time offset of the narrow polystyrene standards and verified by analysing a broad polymer standard.

Results

SEC-Viscometry. Figure 3 illustrates typical DRI raw data chromatograms for a pair of polystyrene narrow standards (Mp=520,000 g/mol and Mp=1,320 g/mol) in three different solvents. The most striking feature of this comparison is the magnitude of the DRI response which of course reflects the specific refractive index increment, v, for each polymer/solvent combination. Assuming a value of v=0.185 cm^3/g for polystyrene in THF (3), values for polystyrene in TCB and CN were interpolated and are summarised in Table II. Results for polystyrene in TCB as a function of temperature are summarised in Table III. The results for the higher molecular weight polystyrene standards in TCB agree well with literature values quoted at similar temperatures (4). The results also show a significant reduction in v for lower molecular weight polystyrene standards which is well documented in the literature (5) and a slight reduction in v for polystyrene generally at higher temperatures. The extremely small value of v for polystyrene in CN implies that for routine calibration procedures using these standards the DRI detector must exhibit very high sensitivity.

Table II. Polystyrene v as a Function of Solvent

Polymer	THF (40°C)	TCB (145°C)	CN (210°C)
Mp 520000 g/mol	0.185	0.051	0.003
Mp 1320 g/mol	0.172	0.033	0.001

units cm^3/g

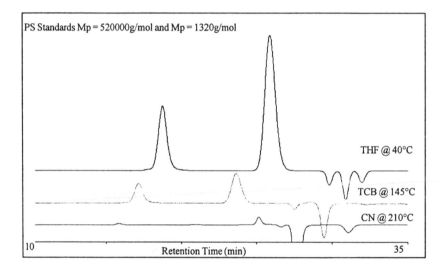

Figure 3.   DRI Detector Response as a Function of Solvent

The peak retention time of the polystyrene standards was also found to vary but since the volumetric flow rate was measured in each case to be precisely 1.0 ml/min, this reduction in retention time with increasing temperature must be associated with changes in the column characteristics.

Table III. Polystyrene ν as a Function of Temperature

Polymer	TCB (145°C)	TCB (160°C)	TCB (180°C)
Mp 210500 g/mol	0.051	0.046	0.040
Mp 580 g/mol	0.015	0.019	0.013

units cm^3/g

From the combined DRI and viscosity detector responses, the intrinsic viscosity, [n], of each of the polystyrene and polyethylene standards was calculated. These values, together with the vendor molecular weight (M) values of the standards, were used to produce a Mark-Houwink-Sakurada plot of log [n] versus log M. According to the Mark-Houwink-Sakurada relationship, $[n]=KM^\alpha$, the slope of this plot equates to α and the intercept to log K. Figure 4 illustrates typical plots for polystyrene in the three solvents studied and Figure 5 shows similar plots for polystyrene and polyethylene in TCB at different temperatures. The K and α values calculated from these plots for polymers with molecular weight greater than 10,000 g/mol are summarised in Table IV. The Mark-Houwink-Sakurada parameters for polystyrene in THF and for polystyrene and polyethylene in TCB at 145°C agree well with literature values (6, 7, 8) and increasing temperature had a relatively small effect on the values determined. The values for polystyrene in CN at 210°C differed somewhat to those reported by Stacey (2) but were fairly consistent with the values determined at 180°C.

Table IV. Mark-Houwink-Sakurada Parameters Determined

Polymer	Experimental K (x10^5)	α	Literature K (x10^5)	α
PS/THF/40°C	13.9	0.714	14.1	0.700
PS/TCB/135°C	12.8	0.690	12.1	0.707
145°C	8.7	0.704		
150°C	9.6	0.690		
160°C	8.3	0.704		
180°C	9.8	0.690		
PE/TCB/135°C	39.0	0.729	40.6	0.725
145°C	32.0	0.746		
150°C	53.0	0.703		
160°C	41.0	0.725		
180°C				
PS/CN/180°C	1.6	0.755	18.6	0.657
210°C	3.3	0.713		
PE/CN/180°C	15.0	0.741	64.0	0.671
210°C	2.4	0.863		

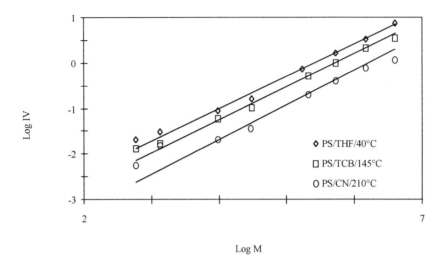

Figure 4. Mark-Houwink-Sakurada Plots for Polystyrene in the Three Different Solvents Studied

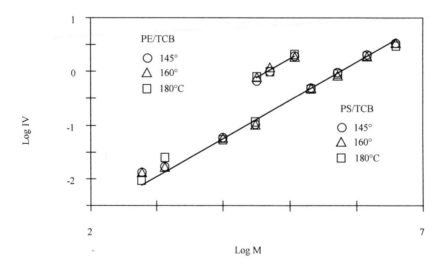

Figure 5. Mark-Houwink-Sakurada Plots for PS and PE in TCB at Different Temperatures

In order to characterize polymers by SEC-viscometry the molecular weight and intrinsic viscosity data for the polystyrene standards was used to produce a Universal Calibration *(9)* where log M[n] was plotted against peak retention time for each standard. Figure 6 shows Universal Calibration plots for polystyrene under three solvent/temperature conditions. In each case the data has been fitted using a first order polynomial which reflects the column characteristics, mixed gel columns are intended to provide a linear conventional log M versus retention time calibration plot. This linear fit is useful as some degree of extrapolation is normally required when employing molecular weight sensitive detectors in order to compensate for the lack of detector sensitivity at the extreme tails of the polymer distribution.

Characterization of Poly(phenylene sulfide) (PPS)

Figure 7 shows typical DRI raw data chromatograms for three PPS samples contrasted with the DRI trace for two polystyrene standards. The PPS samples, prepared nominally at 2 mg/ml, exhibit very good DRI detector response compared to the polystyrene standards and retention time differences indicate variation in molecular weight between the PPS samples, sample A lowest and sample C highest in molecular weight. Typical viscometer raw data chromatograms for the same set of samples is shown in Figure 8. By contrast the polystyrene shows very good response and for the PPS samples both retention time and response height varies in accordance with the molecular weight (and intrinsic viscosity) differences between the samples.

For a set of six PPS samples the weight average molecular weight (Mw) was determined by two methods :

1. Employing DRI response only together with Mark-Houwink-Sakurada parameters for PS and PPS as reported by Stacey *(2)*

2. Employing SEC-viscometry with a Universal Calibration plot generated using polystyrene standards.

The results for the six samples, designated A to F, are summarised in Table V. For the higher molecular weight samples, C through to F, there was reasonably good correlation between the results obtained by the two methods. The two lower molecular weight samples (A and B) exhibited more variation in molecular weight by the two methods. This could be associated with the fact that the Mark-Houwink-Sakurada parameters for PPS reported by Stacey were calculated over a relatively narrow molecular weight range with no data below around 20,000 g/mol.

The molecular weight distributions calculated for the six samples by SEC-viscometry are compared in Figures 9 and 10. The molecular weight differences between samples A, B and C can clearly be seen in Figure 9 but another notable feature was the difference in polydispersity with sample B exhibiting a significant low molecular weight tail. In Figure 10 it can be seen that samples D and E have very similar distributions and that sample F has an overall higher molecular weight but again all

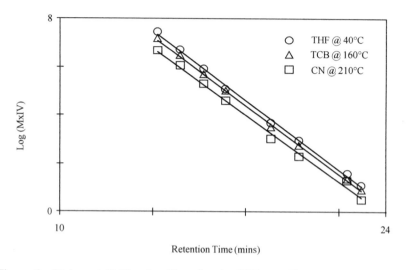

Figure 6. Universal Calibration Plots for the Different Temperatures and Solvents Studied

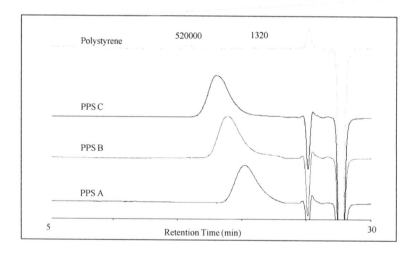

Figure 7. Typical DRI Rawdata Chromatograms of PPS

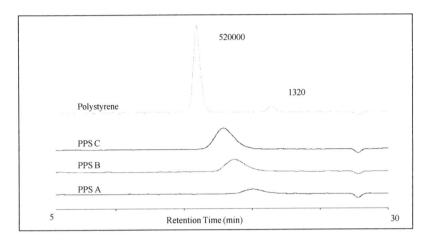

Figure 8. Typical Viscometer Raw Data Chromatograms for PPS

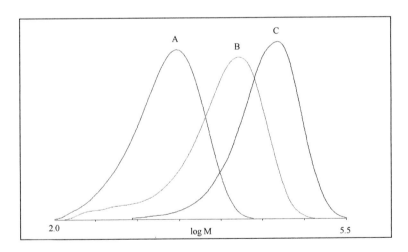

Figure 9. GPC-Viscometry Molecular Weight Distribution Overlays for the PPS Samples A, B and C

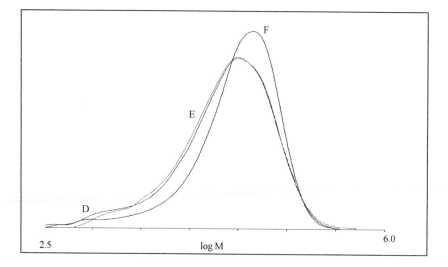

Figure 10. GPC-Viscometry Molecular Weight Distribution Overlays for the PPS Samples D, E and F

three samples show the presence of a low molecular weight tail on the distribution.

Table V. Summary of Results for PPS Samples

Sample	Mw determined by conventional GPC, applying K and α values determined by Stacey	Mw determined by GPC-Viscometry (Universal Calibration generated from PS standards)
A	6851	3100
B	22209	16300
C	51850	47500
D	42422	37000
E	43742	37740
F	46213	42600

SEC-Light Scattering-Viscometry

The results quoted in Table III indicated that as the temperature was increased, v for polystyrene in TCB decreased slightly. In light scattering experiments the scattered light intensity is proportional to the square of the v term and therefore detector sensitivity will be of ultimate concern as the temperature of SEC operation is increased. Work with the triple detector system was more limited than with the SEC-viscometry system but experiments have been carried out to study the feasibility of the technique. Figures 11 and 12 shows the detector responses from the triple detection system for a pair of polystyrene standards (Mp=520,000 g/mol and Mp=9680 g/mol) and for a polyethylene standard (NBS 1475) in TCB at 160°C. The concentrations prepared for these sample solutions were 0.76 mg/ml, 1.72 mg/ml and 2.29 mg/ml. NBS 1475 is quoted as having Mw=52,000 g/mol and for linear polyethylene in TCB at 135°C the published v value is 0.107 cm^3/g (4). The results obtained so far would suggest that the triple detector system is well suited to the characterization of commercial, higher molecular weight polyolefins in TCB at temperatures of 160±20°C. It is also suggested that for PPS characterization in CN at 210°C where, although the polymers have typically low Mw values (20,000-60,000 g/mol), the v is quoted as 0.137 cm^3/g (2), this triple detector system could offer significant advances in the determination of molecular weight distribution.

Conclusions

The Mark-Houwink-Sakurada parameters have been determined for polystyrene and polyethylene in a range of solvents and at various temperatures, some at temperatures well in excess of values previously reported in literature. This work has been performed using a commercial high temperature SEC instrument fitted with DRI and viscometry detectors. The parameters for polystyrene and polyethylene in TCB did not appear to vary significantly over the temperature range 135°C to 180°C. This suggests that similar methodology can be applied routinely for the characterization of polyolefins by SEC-viscometry using higher temperatures.

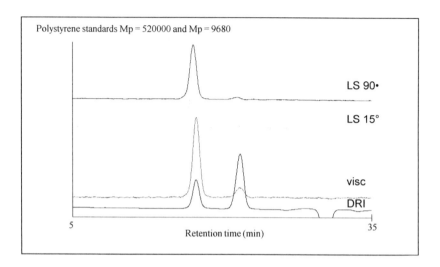

Figure 11. SEC-LS-viscometry in the PL-GPC210

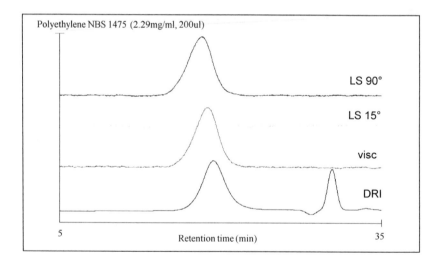

Figure 12. SEC-LS-viscometry in the PL-GPC210

The application of SEC-viscometry to the characterisation of PPS has been shown to be sensitive to changes in both molecular weight and molecular weight distribution. Molecular weight values obtained by this online, multi-detector technique were in good agreement with values calculated using literature quoted parameters which have been generated using off-line techniques.

Based on the work presented here, the DRI response for PPS suggests that the v in CN at 210°C is relatively large and usable for on-line SEC-light scattering experiments. The potential of SEC-LS for the characterisation of PPS will be the subject of future work.

Literature Cited

1. Lesec, J.; Millequant, M. *International GPC Symposium Proceedings*. 1996, 87-115.

2. Stacy, C. J. *J. Appl. Polym. Sci.* **1986**, *32*, 3959-3969.

3. Zigon, M.; The, N. K.; Shuyao, C.; Grubisic-Gallot, Z. *J. Liq Chrom. & Rel. Technol.* **1997**, *20(14)*, 2155-2167.

4. Horska, J.; Stejskal, J.; Kratochvil, P. *J. Appl. Polym. Sci.* **1983**, 28, 3873-3874.

5. Margerison, D.; Bain, D. R.; Kiely, B. *Polymer.* **1973**, *14*, 133-136.

6. Yau, W. W.; Kirkland, J. J.; Bly, D. D. *Modern Size-Exclusion Liquid Chromatography*; John Wiley & Sons: New York, NY, 1979; 337.

7. Lehtenin, A.; Vainikken, R. *International GPC Symposium Proceedings.* **1989** 612

8. Scholte, T.G.; Meijerink, N. L. J.; Schoffeleers, H. M.; Brands, A. M. G. *J. Appl. Polym. Sci.* **1984**, *29*, 3763-3782.

9. Z Grubisic, Z.; Rempp, P.; Benoit, H. *J. Polym. Sci., Part B, Polym. Lett.* **1967**, *5*, 753.

Chapter 6

Use of the Single-Capillary Viscometer Detector, On-Line to a Size Exclusion Chromatography System, with a New Pulse-Free Pump

Raniero Mendichi and Alberto Giacometti Schieroni

Istituto di Chimica delle Macromolecole (CNR), Via Bassini 15, 20133 Milan, Italy

Molar Mass sensitive detector, Light Scattering and Viscometer, on-line to a Size Exclusion Chromatography system presents considerably interest. Unfortunately the commercially available version of the Single-Capillary Viscometer (SCV), that uses a traditional HPLC pump and a pulse dampener, presents serious problems and can give poor performances. Low sensitivity and flow rate fluctuations are the main problems in the use of this on-line SCV detector. In the past several solutions to the problems have been described. Our proposal is to use the SCV detector with a new commercially available pulse-free pump without pulse dampening. A detailed evaluation of this new SEC-SCV system has been performed using various polymers soluble both in organic and in aqueous solvent. Molar mass distribution, intrinsic viscosity distribution and constants of the Mark-Houwink-Sakurada relationship have been determined. The obtained results are very encouraging.

In recent years the attention of the Size Exclusion Chromatography (SEC) users has been focused towards the use of molar mass sensitive detectors, particularly Light Scattering and Viscometer detectors. Since its introduction in 1972 (1), the on-line viscosity-sensitive detector for SEC has become an interesting technique for the determination of Molar Mass Distribution (MMD), Intrinsic Viscosity Distribution (IVD) and constants of the Mark-Houwink-Sakurada (MHS) relationship. In the 80's, two commercial on-line SEC viscometers were introduced. The first one in 1984 (2), was the differential viscometer (DV) from Viscotek. The second one in 1989 (3), was the Single-Capillary Viscometer (SCV) from Waters. The DV detector uses four capillary tubes in a hydraulic Wheatstone bridge configuration and two differential pressure transducers. In the Waters SCV detector a single differential transducer measures the pressure drop across a stainless steel capillary tube. The relative simplicity of the SCV detector has immediately attracted the interest of many SEC users.

While the performances of the DV detector find general agreement, the performances of the SCV detector have been controversial. In literature we can find positive evaluations (3-5), "problematic" evaluations (6) and also poor evaluations (7). There has been a general consensus that the main problem in the use of the SCV detector is flow rate fluctuations, which are a direct consequence of pulse dampening within the chromatographic system. In fact, these fluctuations are the response of the pulse dampener to the change in viscosity of the eluent. To overcome this problem, Lesec has initially proposed to decrease the flow resistance of the polymer solution in the concentration detector (8) using semi-capillary connecting tubes: 0.020" of internal diameter. More recently, Lesec has proposed the use of a second pressure transducer located between the pump and the injector (9). Obviously the second pressure transducer provides flow referencing. In this way Lesec has obtained a new differential viscometer potentially insensitive to the flow fluctuations.

Our idea is the use of the SCV on-line detector with a new commercially available pulse-free HPLC pump without pulse dampening. For six years in our laboratory we have used the conventional Waters SCV detector integrated in the 150CV chromatographic system. In this study we have replaced the pump, dampener and autoinjector of the conventional 150CV system with the pump and autoinjector of the new pulse-free Alliance 2690 pump from Waters. MMD, IVD, and in particular the constants of the MHS relationship have been determined for a detailed evaluation of the new system. This new SEC-SCV system has been evaluated using various polymers soluble both in organic and in aqueous solvent.

Experimental

Materials. To evaluate the new SEC-SCV system we have used several narrow MMD standards and several broad MMD samples from a variety of sources. The polymers were soluble both in organic solvents: Tetrahydrofuran (THF), chloroform and N,N-Dimethylformamide (DMF), and in aqueous solvents. Polymers soluble in organic solvents were: Polystyrene (PS), Poly(methyl methacrilate) (PMMA) Poly(vinyl acetate) (PVAc), Poly(vinyl chloride) (PVC), Poly(vinyl pyrrolidone) (PVP) and Poly(alchylthiofene) (PAT). Polymers soluble in an aqueous solvent were: Poly(ethylene oxide) (PEO), Poly(ethylene glycol) (PEG), Poly(aspart-hydrazide) (PAHy) and linear Polysaccharides as Pullulan and Hyaluronan (HA).

Narrow MMD standards were purchased respectively: PS by Polymer Standards Service (Germany), PEO by Toyo Soda (Japan), Pullulan by Showa Denko (Japan), PEG by Polymer Laboratories (UK). Broad MMD samples were obtained respectively: PS NBS 706 by National Bureau Service (USA), PS Edistir 1380 by Montedison (Italy), PMMA and PVC by BDH (USA), PEO and PAT by Aldrich (USA), PVAc from Mapei (Milan, Italy). PAHy samples were kindly obtained from Prof. Giammona (Palermo, Italy), PVP from Prof. Ferruti (Milan, Italy), HA from Dr. Soltes (S.A.S. Slovak Republic) and from Pharmacia & Upjohn (Nerviano, Italy).

Chromatographic System. An original multi-detector SEC system has been used. Figure 1 shows the schematic of this new system. The system was composed of an Alliance 2690 separations module, pump vacuum degasser and autoinjector, from Waters (Milford, MA, USA) and the SCV and the differential refractometer (DRI) detectors which were integrated in the conventional 150CV chromatographic system from Waters. It is well known that the on-line SCV detector needs of the universal calibration (10). There are many experimental evidences for failure of the universal calibration in highly polar mobile phases. For example in aqueous mobile phase it is common to observe a fractionation of the macromolecules undergoing non-steric behaviour. Therefore, we have used an additional multi-angle laser light scattering (MALS) detector from Wyatt (S. Barbara, CA, USA). In this multi-detector system, SEC-MALS-SCV, the molar mass, by MALS, and the intrinsic viscosity, by SCV, have been measured directly. More detailed description of the SEC-MALS-SCV system and related problems have been reported previously (11-12).

Experimental conditions. According to the mobile phase different experimental conditions were employed. In THF mobile phase the experimental conditions were: 1.0 mL/min of flow rate, 35°C of temperature, a column set composed of four Ultrastyragel columns (10^6-10^5-10^4-10^3) from Waters. In DMF + 0.01M LiBr mobile phase the experimental conditions were: 0.8 mL/min, 50 °C, two Styragel columns (HR4-HR3) from Waters. In chloroform mobile phase the experimental conditions were: 0.6 mL/min, 35 °C, two PLGel Mixed C columns from Polymer Laboratories (UK). Finally, in 0.15M NaCl aqueous mobile phase the experimental conditions were: 0.8 mL/min, 35 °C, a precolumn and two OHpak columns (KB806-KB805) from Shodex (Japan).

Viscometers. The dimensions of the capillary of our viscometer were: 0.014" of internal diameter and 6" of length. Hence, the volume of the detector cell is very low, approximately 18 μL. The full scale of the differential pressure transducer was 5 KPa. Viscometer detector signal depends on the intrinsic viscosity, [η], and on the concentration, c, of the solution. Hence, to obtain constant SCV signal-to-noise ratio the concentration of the samples has been calculated so that to maintain constant the specific viscosity of the sample solutions. Specifically, we have used the rule [η]·c=0.1. This value has been chosen because assures sufficient SCV signal.

For a comparison the intrinsic viscosity of some polymeric samples was also measured in static off-line mode by an Ubbelohde viscometer. Off-line [η] values have been used as reference for SCV on-line values. Off-line viscosity data analysis has been performed by the usual Huggins and Kraemer relationships.

Light Scattering. Light Scattering measurements were performed by an on-line Dawn DSP-F detector from Wyatt. MALS detector uses a He-Ne laser, 632.8 nm of wavelength, and measures the intensity of the scattered light at 18 fixed angular locations ranging, in THF, from 12.3° to 159.7°. MALS hardware and analysis software have been described in detail elsewhere (13) and will not be reported herein. The calibration constant that transform the photodiodes voltage in Rayleigh Factor, R_θ, was calculated using Toluene as standard assuming R_θ=1.406·10^{-5} cm^{-1}. The angular normalization of the photodiodes was performed by a narrow MMD PS

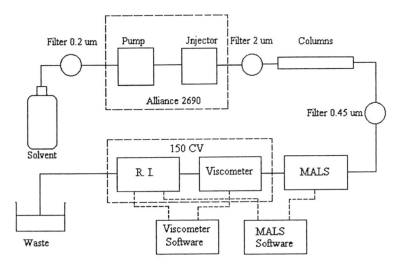

Figure 1.   Schematic of the new SEC-MALS-SCV multi-detector system.

standard, 10.9K of molar mass and D≤1.03, in organic solvent and by a Bovine Serum Albumin globular protein in aqueous solvent assumed to act as isotropic scatterers.

Data Acquisition. SCV data acquisition software was Millennium 2.15 from Waters. MALS data acquisition software was Astra 4.50 from Wyatt. Subsequent to the acquisition, the raw data files have been exported by Millennium and Astra software and elaborated with personal software.

SCV data analysis

Theory, data analysis algorithms, and problems of the SCV detector have been described in detail elsewhere (3-6, 8, 9, 14, 16) and will not be reported herein. Briefly, it will be reviewed only some crucial SCV algorithms. In the SCV data analysis three parameters are fundamental: intrinsic viscosity and concentration at each elution volume (slice) and whole polymer intrinsic viscosity.

In the on-line SCV characterization the extrapolation to infinite dilution is not possible. The intrinsic viscosity at each elution volume, $[\eta]_i$, have to be calculated from a single concentration data. Hence, to estimate $[\eta]_i$ we have used the Solomon-Ciuta equation (15).

$$[\eta]_i = \frac{\sqrt{2}}{c_i} \cdot \sqrt{\eta_{sp,i} - \ln(\eta_{r,i})} \quad (1)$$

Where the subscript i denotes the ith slice of elution volume V_i, c the concentration, η_{sp} and η_r respectively the specific viscosity and the relative viscosity of the solution.

To estimate the concentration at each elution volume, c_i, the equation 2a has proven more reliable with respect to the equation 2b. In fact, the equation 2b assumes that the recovered mass, calculated from the whole area of the DRI chromatogram, matches the injected mass of the polymer. It is well known that, particularly in aqueous solvent, equation 2b can lead to significant error in the value of the concentration of the slices (16). In fact, very often in aqueous solvent the recovered mass, $W_{Rec}=\Sigma w_i=\Sigma c_i \cdot \Delta V$, is significantly lower than the injected mass. The explanation is not simple, depends on the polymer and the mobile phase used. Sometimes absorption of the polymer can occur. Frequently, the polymeric sample, in particular biopolymers, contains low molar mass impurities as solvents, water, etc.

$$c_i = \frac{K_{RI}}{dn/dc} \cdot H_i \quad (2a) \qquad c_i = \frac{W_{inj}}{\Delta V} \cdot \frac{H_i}{\sum H_i} \quad (2b)$$

Where W_{Rec} denotes the recovered mass of the polymer, w_i the mass in the slice, ΔV the elution volume between two acquisition points, H_i the height of the signal of the DRI concentration detector subtracted of the baseline, K_{RI} the calibration constant of the DRI detector, and dn/dc the specific refractive index increment of the polymer. The K_{RI} calibration constant can be determined by injecting a sample with known dn/dc value into the DRI detector at a few different concentrations.

Finally, to estimate the whole polymer intrinsic viscosity, [η], the equation 3a, that uses only the viscometer signal and the injected mass of the polymer, has proven more reliable with respect to the equation 3b that uses the signals of both two detectors SCV and DRI. The derivation of the equations 3a and 3b can be found in the reference (4).

$$[\eta] = \frac{\Delta V}{W_{inj}} \cdot \sum \frac{P_i - P_o}{P_o} \quad (3a) \qquad [\eta] = \frac{\sum c_i \cdot [\eta]_i}{\sum c_i} \quad (3b)$$

Where W_{inj} denotes the injected mass of the polymer, P_i and P_o respectively the signal of the pressure transducer due to the polymer solution and to the pure mobile phase (baseline). In conclusion, to estimate the parameters $[\eta]_i$, c_i, and $[\eta]$ our software uses respectively the equations 1, 2a, and 3a.

Results and Discussion

According to the Hagen-Poiseuille law, equation 4, the pressure drop, ΔP, across the capillary viscometer, SCV signal, depend on the flow rate, F, and on the viscosity of the solution, η.

$$\Delta P = \frac{8 \cdot L \cdot F \cdot \eta}{\pi \cdot R^4} = K_v \cdot F \cdot \eta \quad (4)$$

Where L and R denote respectively the length and internal diameter of the capillary and K_v the viscometer constant. Specifically, the SCV signal is highly flow rate sensitive. For this reason, the key element of the commercial SCV detector, integrated in the commercial Waters 150CV system, is the so-called Baseline Optimization Box (BOB). The BOB consists of a series of eight dampeners and restrictors in alternating order that reduce the flow fluctuations, intrinsic to a reciprocating dual piston pump, by a factor of more than 100 (3). In reality the baseline due to the pure mobile phase of the on-line SCV detector that uses the BOB dampener is very stable as long as column frits are clean. Problems arise when the polymeric solution comes across the chromatographic system. BOB behaves as a hydraulic "capacitance" (6) and when the device compensates for additional resistance secondary, yet variable, flow fluctuations occur. Every time the polymeric solution meets an "obstacle" (frits of the columns, detector cell, etc.) flow rate initially decrease then increases creating a cycling pattern (6). As consequence, the viscometer peak shifts to later retention volumes and the estimated slope of the MHS equation using a single broad MMD sample is underestimated. This phenomenon is particularly prevalent when the polymeric solution flows through a high resistance refractometer. This phenomenon, "Lesec effect", is well known and has extensively studied in the past (4-6). To overcome the flow fluctuation problem we suggest using the SCV on-line detector with a new pulse-free pump without adding additional dampening.

Signal of the new SCV detector. At constant mobile phase viscosity, the SCV detector is a very sensitive flowmeter. Figure 2 shows the signals, SCV and DRI, of

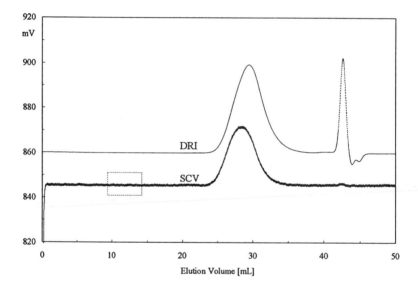

Figure 2.   Signals, SCV and DRI, of the new SEC-SCV system: broad MMD PMMA sample, THF mobile phase, 0.8 mL/min, 25 °C.

the new SEC-SCV system without dampener. The Figure shows the raw SCV signal without any treatment. Figure 3 shows an enlarged portion of the baseline of the new SCV detector. The experimental conditions were: THF, 0.8 mL/min, 35 °C. We can see a residual oscillation, noise, with a period of about 3-4 seconds. The amplitudes of these oscillations depend on the flow rate and a little on the solvent: approximately 1.2 mV in THF, 1.1 mV in chloroform, 1.6 mV in DMF and 1.8 mV in water.

Treatment of the noise of the SCV signal: It is worthy to note that the oscillations (noise), period and amplitude, of the SCV signal are very constant. For this reason it is relatively simple to eliminate it. For example we can use a Fast Fourier Transformate (FFT) filter as proposed from Provder et al. (14). The frequency signal spectrum, for a broad MMD PS sample Edistir 1380, is shown in Figure 4. Instead, Figure 5 shows the signals before and after the FFT filter. For convenience the signal before the FFT filter has stayed shifted of 4 mV. In the FFT filtration we have set to zero the amplitude of the frequency between 0.7 and 2π-0.7.

An alternative to the FFT filter consists in the use of the Savintsky-Golay smoothing algorithm (17) implemented in the Millennium software. In the elaboration of the following results we have used the Savintsky-Golay algorithm, 21 points (21 seconds). However, the final smoothed chromatograms obtained by the Savintsky-Golay algorithm were very similar to that obtained by the FFT digital filter. After the digital filter the noise decreases of a factor about 10, the signal-to-noise ratio increase consequently and very important area and height of the chromatogram remain unchanged. Table I reports a comparison between the results of the FFT digital filter and of the Savintsky-Golay smoothing algorithm on the SCV signal of a broad MMD PS sample in THF solvent. Similarly, a comparison on the SCV signal of some narrow MMD standards does not show substantial difference between the two digital filters. Hence, although the FFT filter is more flexible, there is not substantial difference between the two methods.

Table I. Comparison between the FFT digital filter and the Savintsky-Golay smoothing algorithm: PS sample in THF.

	Raw File	**FFT**	**Savintsky Golay**
Noise (mV)	1.2	0.11	0.12
Signal/Noise	20	218	210
Area (μV·s)	3,472,200	3,468,125	3,465,780
Height (μV)	15,905	15,877	15,868

Interdetectors delay volume. An important problem in the use of a multi-detector SEC system is the alignment of the signals of the different detectors. The importance of the matter has been attested by the volume of publications that have issued. Specifically, a correct value of the interdetectors SCV-DRI delay volume must be accounted for. It is well known that the value of the interdetectors delay volume

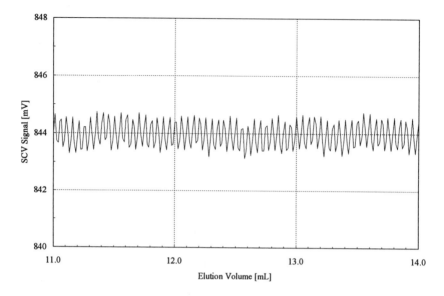

Figure 3.   Noise of the new SCV detector: THF mobile phase, 0.8 mL/min, 35 °C.

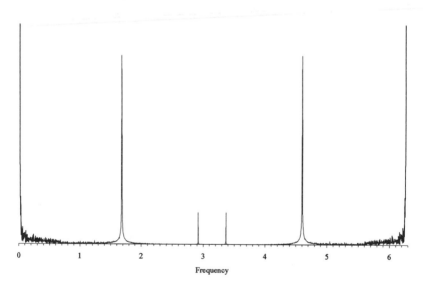

Figure 4.   Frequency signal spectrum of the new SCV detector for a broad MMD PS sample.

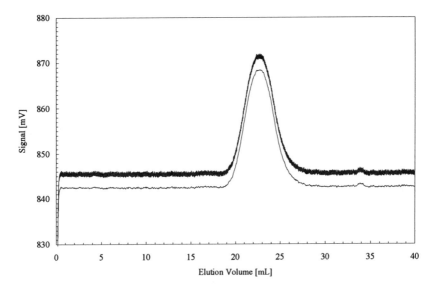

Figure 5. SCV detector signal before and after the FFT filtration for a broad MMD PS sample.

influences notably the values of the MHS constants recovered from a single broad MMD sample (4, 5, 8, 14, 21). The value of the interdetectors delay volume that we have used in the data reduction software has been estimated by a numerical optimization of the best superimposition between the experimental $[\eta]=f(V)$ SCV calibration, from a broad MMD linear sample, and the classical $[\eta]=f(V)$ calibration from some narrow MMD standards. Our method is similar to the method reported by Balke et al. (21). Figure 6 shows the superimposition between the experimental $[\eta]=f(V)$ calibration, from a broad MMD PS sample, and the classical calibration from some narrow MMD PS standards. The delay volume value used in our data reduction software, 80 µL, was approximately equal to the calculated physical delay volume, connecting tubes plus detectors cell, of the system. We retain that this datum is very important. In fact, in the "conventional" 150CV system to obtain reliable MHS constants from a single broad MMD sample it is necessary to use a lower value of the delay volume, and in some cases a negative value, to compensate for the flow fluctuation problem (8).

Evaluation of the new SEC-SCV system. Intrinsic viscosity, MMD and the constants of the MHS relationship have been used to verify the accuracy of the new SEC-SCV system.

    Accuracy of the new SEC-SCV system. Table II reports a comparison between the whole polymer intrinsic viscosity, $[\eta]$, results estimated by the new on-line SCV detector and by a conventional off-line Ubbelohde viscometer on some broad MMD samples. Here and in the following $[\eta]$ values are expressed in dL/g. Table II reports some results for broad MMD PEO and HA samples in aqueous solvent and for broad MMD PS samples in organic solvent. The agreement between on-line and off-line $[\eta]$ data was very good. In fact, the difference between on-line and off-line $[\eta]$ value was lower than 0.6%.

Table II. Comparison between $[\eta]$ results estimated by the new on-line SCV detector and by an off-line Ubbelohde viscometer.

Sample	Solvent	Mw	D	$[\eta]$ [1]	$[\eta]$
		g/mol		dL/g	dL/g
PEO Aldrich	0.15M NaCl	114,600	3.66	1.005	1.012 [2]
PEO Aldrich	0.15M NaCl	434,700	3.84	3.040	3.058 [2]
HA Soltes	0.15M NaCl	660,000	2.04	11.83	11.88 [3]
PS Edistir 1380	THF	230,700	2.38	0.903	0.892 [2]
PS NBS 706	THF	267,600	2.45	0.941	0.936 [2]

(1) On-line SEC-SCV viscometer; (2) Off-line Ubbelohde viscometer; (3) Rotational viscometer

    Furthermore, Table II reports the weight average molar mass, Mw, and the polydispersity, D, of five broad MMD samples soluble in 0.15M NaCl or in THF. Mw and D results match very well the expected nominal values of the five samples.

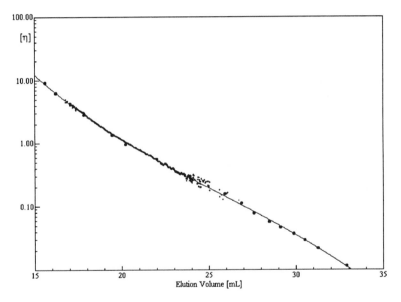

Figure 6. Superimposition between the experimental $[\eta] = f(V)$ calibration from a broad PS sample and the calibration from some PS narrow standards.

With "nominal value" we mean the absolute value reported by the vendor of the polymer or, if missing, the value estimated by an independent light scattering characterization.

Narrow MMD standards characterization: Table III summarizes some results for [η] estimated by the new SEC-SCV system for PEO, PEG and Pullulan narrow MMD standards in aqueous 0.15M NaCl mobile phase at 35 °C. The molar mass of the PEO standards ranges from 860K to 21K g/mol. The molar mass of the PEG standards ranges from 23K to 600 g/mol. Finally, the molar mass of the Pullulan polysaccharides standards ranges from 853K to 5.8K g/mol.

Our [η] data for PEO, PEG and Pullulan narrow MMD standards are in good agreement with the expected data reported in literature (18, 19) for the polymers. Besides [η] values span a broad range inclusive of very low values. These results confirm that the low sensitivity of the SCV detector is not an insurmountable problem.

Table III. Intrinsic viscosity of some narrow MMD SEC standards in 0.15M NaCl aqueous solvent at 35 °C.

PEO		PEG		Pullulan		PS	THF	CHCl₃
Mp g/mol	[η] dL/g	Mp g/mol	[η] dL/g	Mp g/mol	[η] dL/g	Mp g/mol	[η] dL/g	[η] dL/g
860,000	4.200	23,000	0.333	853,000	1.649	5,480,000	8.160	-
570,000	3.027	12,600	0.252	380,000	0.983	900,000	2.298	2.450
270,000	2.023	7,100	0.178	186,000	0.610	225,300	0.871	0.999
160,000	1.287	4,100	0.122	100,000	0.404	105,560	0.525	0.520
85,000	0.835	1,470	0.069	48,000	0.256	43,000	0.267	0.280
45,000	0.538	960	0.056	23,700	0.156	10,900	0.099	0.098
21,000	0.319	600	0.045	12,200	0.104	4,016	0.050	0.046
				5,800	0.066	1,060	0.032	0.031

Furthermore, Table III reports some [η] data for narrow MMD PS standards both in THF and in chloroform solvent. The molar mass of the PS standards ranges from 5.48M to 1060 g/mol. Our [η] data for PS standards are congruent with the [η] values calculated by the MHS relationship and the constants reported in literature (4, 20) for PS polymer in THF and chloroform solvent.

From [η] and Mp data of the narrow MMD standards, summarized in the Tables III, we can compute the MHS constants for PEO, PEG and Pullulan polymers in 0.15M NaCl solvent and for PS polymer in THF and chloroform solvent. The recovered MHS constants were in good agreement with the expected values reported in literature for the polymers. The recovered values of the MHS constants have been reported elsewhere (22).

Broad MMD samples characterization: Figure 7 shows the experimental $[\eta]=f(V)$ calibration, where V denotes the elution volume, estimated by the new SEC-SCV system with a broad MMD PS sample NBS 706. The signal-to-noise ratio was very good and with the exception of the extremities of the chromatogram the local $[\eta]_i$ values were very accurate. Figure 8 shows the relative MHS plot, in THF solvent at 35 °C. Disregarding the extremities of the plot where the signal-to-noise ratio was poor we have recovered the following MHS constants: $k=1.40 \cdot 10^{-4}$ and $a=0.708$. These values were in good agreement with the expected values reported for PS polymer in THF solvent (4, 20).

To estimate the MHS constants we have used a simple, non-sophisticated, algorithm. The "good data region" has been estimated using the "threshold method" (4). The threshold value defines the data points, namely the "good data region", used in the $\text{Log}([\eta])=f(\text{Log}(M))$ fitting. Specifically, both for DRI and SCV chromatograms, the threshold value was 5% of the peak height. Besides, we have used direct, non-weighted, linear fitting.

Table IV reports a review of the constants of the MHS relationship recovered by the new on-line SCV detector for many polymers in different solvents. Some characterizations concern polymers quite usual for SEC for which exists large data in literature. In this case the comparison between the obtained and the expected results is meaningful. Some other characterizations concern new polymers for which do not exist data in literature. Some of these last characterizations, for different reasons, are critical. However, these results were in good agreement with the values reported in literature (14, 18-23).

Table IV. Summary of the Mark-Houwink-Sakurada constants recovered with the new on-line SEC-SCV system and broad MMD samples

Polymer	Sample	Solvent	$k \cdot 10^4$	a
PS	NBS 706	THF	1.40	0.708
PMMA	BDH	THF	1.14	0.699
PVAc	Mapei	THF	1.01	0.760
PVC	BDH	THF	3.96	0.701
PAT	Aldrich	THF	1.20	0.727
PS	Edistir 1380	CHCl₃	0.92	0.752
PVP	Ferruti	DMF + 0.01M LiBr, 50 °C	6.10	0.556
PEO	Aldrich	0.15M NaCl	3.94	0.699
HA	Soltes	0.15M NaCl	3.67	0.785
PAHy	Giammona	0.1M NaNO₃ + P.B. pH 7.8	3.44	0.531

PAT sample was a commercially available conducting polymer, Poly(octyl-thiofene), from Aldrich and the recovered MHS constants, $k=1.20 \cdot 10^{-4}$ and $a=0.727$,

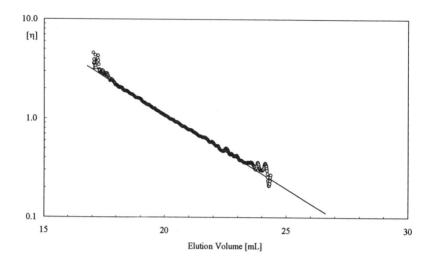

Figure 7. Experimental $[\eta] = f(V)$ calibration of the new SEC-SCV system for a broad MMD PS sample.

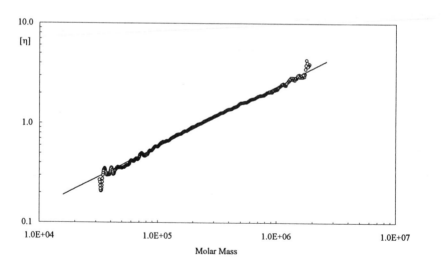

Figure 8. Mark-Houwink-Sakurada plot, $[\eta] = f(M)$, for PS polymer in THF solvent.

were closely to that recovered on a previous study using a different, not commercially available, PAT samples (24).

A critical characterization regards the PAHy polymer in 0.1M NaNO₃ + 0.05M Phosphate Buffer pH 7.8, mobile phase at 35 °C. This complex mobile phase has been studied to eliminate aggregation and interaction of the macromolecules with the packing of the columns. Intrinsic viscosity and molar mass of the sample were very low: $[\eta]$ = 0.065 dL/g, Mw = 21,800 g/mol, D = 2.8. It is well known that the sensitivity of the SCV detector is generally less than the DV detector. Despite this unfavourable situation, choosing an appropriate concentration of the sample (approximately 0.5%), we have recovered values of the MHS constants, k=3.44·10⁻⁴ and a=0.531, which are consistent with these low molar mass samples. Figure 9 shows the MHS plot for PAHy polymer in the previous cited solvent at 35 °C.

Finally, Figure 10 shows the experimental MHS plot for a branched PVAc sample in THF solvent at 35 °C. The MHS constants, k=1.01·10⁻⁴ and a=0.76, for PVAc polymer have been calculated from the linear portion, M ≤ 9.2·10⁴ g/mol, of the $[\eta]=f(M)$ plot. Again these values were in good agreement with the expected values reported for PVAc branched polymer in THF solvent (14).

Conclusions

To resolve the problems of the commercially available version of the SCV detector we have proposed a new SEC-SCV system. Hence an original multi-detector SEC-MALS-SCV system composed of an Alliance 2690 pulse-free pump, without dampener, SCV, MALS and DRI detectors has been tested. A number of narrow MMD standards and broad MMD samples both in organic and in aqueous solvent have been used for a detailed evaluation of the new SEC-SCV system.

The new on-line viscometer, without dampener, shows a residual noise that is easily eliminated by means of a digital filter. The new system provides accurate intrinsic viscosity and MMD results. The MHS constants obtained from this system, using both narrow MMD standards and broad MMD samples, are very closely to the reported values in literature. The interdetectors delay volume, between the SCV and the concentration detector, estimated by an optimization method, was approximately equal to the physical interdetectors delay volume of the system. This result confirms that the flow rate upset problem, typical of a "conventional" SEC-SCV system, has been resolved, or at least minimized. Results obtained with this new SCV system are very encouraging.

References

1. Ouano, A. C.; *J. Polym. Sci. A1*, **1972**, *10*, 2169.
2. Haney, M. A.; *Am. Lab.*, **1985**, *17*, 41.
3. Ekmanis, J. L.; *Proc. Int. GPC Symp.*, Newton MA, **1989**, 1.
4. Kuo, C. Y.; Provder, T.; Koehler, M. E.; *J. Liq. Chromatogr.*, **1990**, *13*, 3177.
5. Lesec, J.; Millequant, M.; Haward, T.; *Proc. Int. GPC Symp.*, S.Francisco CA, **1991**, 285.

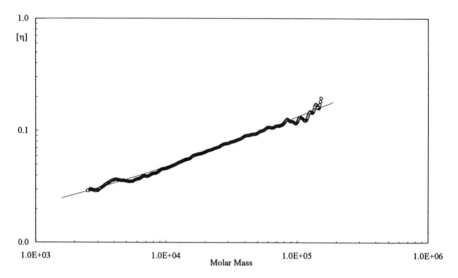

Figure 9. Mark-Houwink-Sakurada plot for PAHy polymer in 0.1M NaNO₃ + 0.05M Phosphate Buffer pH 7.8 solvent.

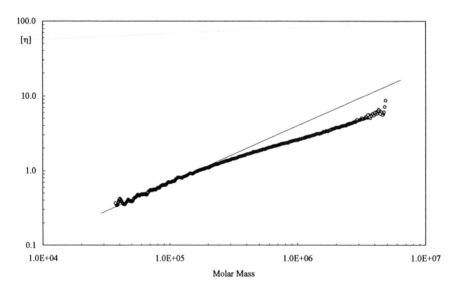

Figure 10. Mark-Houwink-Sakurada plot for PVAc polymer in THF solvent.

6. Yau, W. W.; Jackson, C.; Barth, H. G.; *Proc. Int. GPC Symp.*, S. Francisco CA, **1991**, 263.
7. Dobkowski, Z.; Cholinska, M.; *Proc. 10th Int. Conf. on Macromol.*, Bratislava SK, **1995**, 72.
8. Lesec, J.; Millequant, M.; *J. Liq. Chromatogr.*, **1994**, *17*, 1011.
9. Lesec, J.; Millequant, M.; *Proc. Int. GPC Symp.*, S. Diego CA, **1996**, 87.
10. Grubisic, Z.; Rempp, P.; Benoit, H.; *J. Polym. Sci. B*, **1967**, *5*, 753.
11. Mendichi, R.; Maffei Facino, R.; Audisio, G.; Carini, M.; Giacometti Schieroni, A.; Saibene, L.; *Int. J. Polym. Anal. & Charact.*, **1995**, *1*, 365.
12. Mendichi, R.; Giacometti Schieroni, A.; *Proc. 9th ISPAC, Oxford UK*, **1996**, B22.
13. Wyatt, P. J.; *Anal. Chim. Acta*, **1993**, *272*, 1.
14. Kuo, C. Y.; Provder, T.; Koehler, M. E.; Kah, A. F.; *ACS Symp. Ser.*, **1987**, *352*, 130.
15. Solomon, O. F.; Ciuta, I. Z.; *J. Appl. Polym. Sci.*, **1962**, *6*, 683.
16. Reed, W. F.; *Macromol. Chem Phys.*, **1995**, *196*, 1539.
17. Savitsky, A. A.; Golay, M. J. E.; *Anal. Chem.*, **1964**, *36*, 1627.
18. Nagy, D. J.; *J. Liq. Chromatogr.*, **1990**, *13*, 677.
19. Kawaguchi, S.; Imai, G.; Suzuki, J.; Miyahara, A.; Kitano, T.; Ito, K.; *Polymer*, **1997**, *38*, 2885.
20. Kurata, M.; Tsunashima, Y.; *Polymer Handbook*, 3rd Ed., John Wiley and Sons, New York, N. Y., **1989**, Chap. VII, 1.
21. Cheung, P.; Balke, S.T.; Mourey, T. H.; *J. Liq. Chromatogr.*, **1992**, *15*, 39.
22. Mendichi, R.; Giacometti Schieroni,A.; *J. Appl. Polym. Sci.*, submited.
23. Bothner, H.; Waaler, T.; Wik, O.; *Int. J. Biol. Macromol.*, **1988**, *10*, 287.
24. Mendichi, R.; Bolognesi, A.; Geng, Z.; Giacometti Schieroni, A.; *Proc. Int. GPC Symp.*, Orlando FL, **1994**, 827.

Chapter 7

Molecular Weight Characterization of Polymers by Combined GPC and MALDI TOF Mass Spectroscopy

J. L. Dwyer

Lab Connections, Inc., 201 Forest Street, Marlborough, MA 01752

MALDI TOF MS is a valuable tool for polymer characterization. It is used in the analysis of low molecular weight oligomers, and in determination of polymerization mechanisms, but has limitations in application to high molecular weight species. A high polydispersity sample will yield an erroneous molecular weight distribution by MALDI. GPC fractionation plus MALDI overcomes this problem, but is labor intensive and complex.

Sample preparation methods are also troublesome. Solute-plus-matrix deposited from solution gives rise to segregation of components. The spectroscopist must hunt for "hot spots" in the sample, and anticipate variances in analysis of polymer blends.

Novel sample processing technology and apparatus for MALDI TOF polymer analysis is described. GPC column eluant flows through a nebulizing nozzle, delivering a focused spray which is deposited on a pre-coated matrix plate. The plate transits under the nozzle as chromatography proceeds, resulting in a track of solute/matrix mixture. Any point on this track has a very low polydispersity; and when the matrix plate is traversed through the sample chamber of the MALDI-TOF spectrometer, one obtains a series of molecular weight spectra; which are recombined to provide an accurate molecular weight distribution. All operations are automated, and the system is capable of unattended multi-sample processing.

Examples are presented of analysis of polymer samples, detailing experimental procedure, and techniques of calculation.

Matrix Absorbed Laser Desorption Ionization Time of Flight (MALDI - TOF, MALDI-MS) Mass Spectroscopy is proving to be a very useful tool for the analysis of polymeric materials. MALDI is characterized as a soft ionization method that produces predominantly singly charged molecules with a minimum of fragmentation.

The technique also has the ability to vaporize and ionize large molecules. These characteristics have made the technique especially attractive to the polymer analyst. Briefly, MALDI spectroscopy is obtained by mixing with an analyte a matrix chemical, typically an organic acid. When this mixture is irradiated by a laser beam, it is vaporized, and the analyte aquires an ion charge. The charged analyte molecules are then accelerated through the flight tube and impact the detector in at times characteristic of their mass/charge ratio.

MALDI Sample Preparation Issues

It was initially hoped that MALDI-MS would provide a one-step procedure for polymer molecular weight analysis, eliminating the need for Gel Permeation Chromatography (GPC) fractionation. The need for a series of polymer calibration standards would similarly be eliminated. As MALDI-MS was applied to polymer applications, it became apparent that the response characteristics of the detectors employed posed a problem with broad polydispersity samples typical of polymers; in that sensitivity is biased to the low molecular weight portion of the sample. Although the MALDI spectrum reflects all molecular weight molecules in the sample, the distribution of weight is measurably biased toward the low molecular weight region.

This problem can be addressed by using a combination of GPC and MALDI-MS. The chromatograph eluant at any point in time is of narrow polydispersity, and MALDI provides a well-resolved accurate measurement of the population. The preparation methods used for MALDI-MS samples have presented some issues in the analysis of polymers. The classical method of sample preparation is to co-dissolve the sample and matrix reagent, place a small drop of this preparation on a metal target, evaporate the solvent, and analyze the dried residue in the spectrometer. This preparation is typically uneven, with most of the sample depositing at the outer perimeter of the original droplet (see Figure 1). During evaporation polymer sample may segregate from the matrix chemical, due to differing solubilities of these two species. In polymer blend samples, this same segregation can be observed for the different polymer species present in the sample.

The net result is a spatially non-uniform sample preparation. The spectroscopist must hunt over the area of the sample deposit to find "hot spots" that yield strong spectra. Composition can vary as laser shots are taken at these various positions, with the result being variation in the spectra obtained.

Our laboratory has been working on equipment and techniques that provide a direct coupling of GPC with MALDI-MS. This technology also addresses the issue of sample preparation variability. The coupling of GPC and MALDI sample preparation is a solution to the limitation of polydispersity in polymer MALDI-MS, in that column eluant at any point in the chromatogram is sufficiently monodisperse to provide accurate measurement of the eluant molecular weight distribution. Described below is the GPC-MALDI methodology, and several examples of polymer analyses.

Pre-formed Matrix Targets

Rather than co-depositing matrix and sample, we have developed a method of pre-formation of a matrix coating onto which polymer sample is deposited. A thin sheet of stainless steel is coated with a uniform layer of a matrix chemical of choice; dihydroxy-

86

benzoic acid for example. This coating is 200 - 300 microns in thickness. Inspection with the electron microscope reveals a uniform non-crystalline structure of irregular granules, approximately 5 microns in diameter. This structure is moderately porous When a small amount (1 μl) of a dissolved solute is applied to this matrix coating, the solvent will partially dissolve the top layer of matrix. Because of the porosity of the matrix coating, the droplet is readily imbibed into the coating. Matrix coating is in excess, so that the solvent rapidly saturates with matrix, and only the top surface of the coating is dissolved. Upon evaporation, an electron micrograph reveals the formation of microcrystals from the matrix granules, predominantly in the topmost layer of matrix. The solute is evenly distributed in this microcrystalline crust, and is well included with the matrix chemical.

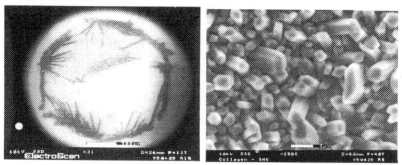

Figure 1Electron micrograph of conventional dried drop preparation (left) and matrix film(right) after sample spray application.

When such a preparation is placed in a spectrometer and mapped with the laser, it can be seen that all regions within the original sample drop boundary produce approximately the same intensity of spectra. There is no requirement to generate a search over the sample deposit area to find hot spots that will yield a spectrum. We are able to produce such precoated targets with a variety of matrix agents, including dihydroxy benzoic acid (DHB), α-cyano-4-hydroxycinnamic acid (α–CCA), 3-hydroxypicolinic acid (HPA), 3-β-indoleacrylic acid (ILA), 2-(4-hydroxyphenylazo)-benzoic acid (HABA), sinapinic acid, and dithranol.

The GPC - MALDI Interface

This precoated target technology has been combined with a nozzle spray deposition system to form a GPC-MALDI MS interface. Eluant from the GPC unit flows to a capillary spray nozzle positioned above a matrix coated target, as shown in Figure 2. The target is placed on a 2-axis automated stage that moves continuously during the chromatogram. As a result the chromatogram is transformed into a continuous stripe of eluant, spray deposited onto the matrix target. Stage movement can be programmed by the operator, such that the deposition pattern conforms to the sample spatial requirements of the MALDI spectrometer being used.

The nozzle utilizes a length of stainless steel capillary tube to deliver and nebulize the chromatograph eluant. This capillary tube is encased in a heated chamber..

Preheated gas (sheath gas) flows into this chamber, across the surface of the capillary, and exits the chamber at the tip of the spray head. This sheath gas does three things:
1. The sheath gas flowing past the nozzle tip provides nebulization of the emerging liquid stream.
2. It provides thermal energy to evaporate the chromatography solvent
3. It serves to contain the nebulized spray in a tightly focused, small diameter emergent cone.

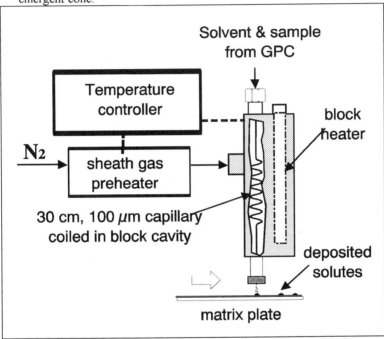

Figure 2 Functional diagram of the GPC-MALDI interface.

Decreasing the diameter of a nebulizer delivery tube of a nozzle produces a reduction in spray pattern diameter. Conventional spray systems can deliver a very small diameter spray; but have very limited flow capacity. Increasing the flow beyond some limit (dependent on solvent composition) results in unstable operation, and insufficient evaporative capacity. It is imperative that almost all of the solvent is evaporated before the spray impacts that matrix coated target. If not, the spray will wash away the matrix coating.

The Linear Capillary Nozzle (patented) is based on the discovery that a relatively long length of capillary can greatly increase the solvent evaporative capacity. The capillary is made of thin-walled stainless steel. This provides superior heat transfer properties. Liquid enters the upstream end of the capillary at a pressure considerably above ambient, because of the flow resistance of the capillary. The pressure decreases linearly along the capillary, to ambient level at the discharge end. As liquid flows through the capillary, it absorbs heat from the hot flowing gas surrounding the capillary. Because of the small dimensions the [capillary surface / liquid volume] ratio is high, and the capillary provides an efficient conduit for the heat required to vaporize the flowing solvent.

The result of this design are opposing gradients of liquid pressure and temperature along the length of the capillary. A fluid element traveling through the capillary is simultaneously rising in temperature and falling in pressure. The flowing liquid flashes to vapor at a point very near the exit end of the capillary. This results in greatly increased liquid processing capacity over what can be obtained with a short capillary, and eliminates entirely the need for costly vacuum environments. The nozzle has been applied to a broad range of chromatography solvents. It can process 100% water and high boiling point organics.

Control of the stage motion and nozzle operating parameters is effected via special Graphical User Interface software operating in a Windows environment. After deposition of the chromatogram on the matrix target, the target is transferred to the sample holder of the MALDI spectrometer and scanned. Because the need to search for hot spots has been eliminated, it is possible to program the MALDI stage to rapidly scan the length of the deposition track, acquiring a total ion current chromatogram. The operator can then go back to regions of maximum interest and collect/coadd spectra for analysis.

Comparison of Dried Drop vs. Nozzle Spray Sample Repeatability

The issues cited with traditional dried droplet sample preps. are familiar to mass spectroscopists. We conducted an experiment to quantitatively compare sample consistency with the dried drop and the spray deposition method. A mixture of polyethylene glycol (PEG) oligomers was used as the sample. Sample and matrix were codissolved as shown:

Sample preparation	
Matrix	2,5-dihydroxybenzoic acid (DHB), 1% in ACN
Sample	PEG 1500 /PEG 4600; 1% ea in MeOH
Matrix / sample ratio	18 / 1
Carrier solvent	50/50 MeOH:H$_2$O

Two μl samples were applied to targets. For the dried droplet measurement, the sample was applied to a clean stainless steel target in a target ring 1.5 mm in diameter. For the nozzle injection experiment, a steady stream of carrier solvent was delivered from a syringe pump, through an HPLC injection valve, and to the nozzle.

Sample Application Conditions	
Sample volume	2 μl
Carrier solvent volume	40 μl
Carrier solvent flow rate	50 μl / min
Nozzle gas pressure	40 psi
Nozzle gas temperature	95°C

The droplet sample, after solvent evaporation, appeared as a white ring of matrix crystals, with little or no material evident in the center of the ring. Optical microscopic examination of the ring area showed crystals of DHB 100 - 200 microns in length. The spray deposit was a compact uniform circular deposit ~ 1.5 mm in diameter. Low power microscopy showed very small crystals uniformly distributed within the deposit circle.

Samples were analyzed by ThermoBioAnalysis on a Vision 2000 reflectron instrument (ThermoBioAnalysis Ltd.). For each deposit 50 spectra were acquired from randomly chosen locations, each spectrum representing the sum of 30 laser shots. Figure 3 shows one such spectrum. Since the two polymers in the sample were present in a 1:1 mass ratio, intensity differences are expected due to the ca. 3/1 difference molar ratios. As can be seen qualitatively, the population here is biased toward the low molecular weight component in this unfractionated sample.

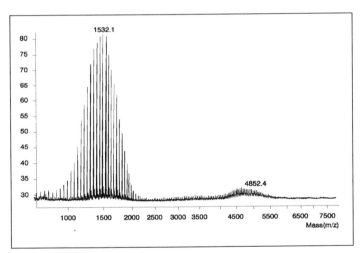

Figure 3 Mass spectrum of PEG sample blend. The numbers indicate maximum intensity in the two populations.

It was noted that the laser threshold was quite consistent across all areas of the sprayed deposits, while the threshold for the conventional deposits depended strongly on the crystal morphology. As a consequence the data acquisition from the sprayed deposits was quicker, with little required operator judgment. Figure 4 shows plots by the two methods of the mass ratio of the two PEG components for 50 samples.

It is apparent that dried droplet method produces significantly more data scatter. This is apparently due to segregation of the PEG 1500 and PEG 4600 during the solvent evaporation and formation of analyte / matrix precipitates. In the spray process there is simply no opportunity for the components to segregate, and this results in much better shot-to-shot repeatability.

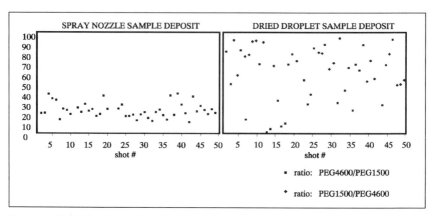

Figure 4 Calculated mass ratio of sample, taken at 50 deposit locations.

GPC - MALDI Experiments

The same sample system (1:1 PEG 1500 / PEG 4600) was then used in a GPC - MALDI experiment. MALDI targets were prepared by pre-coating adhesive backed stainless steel foils with the DHB matrix. DHB was applied as a series of parallel tracks 3 mm wide by 50mm long, and the coating thickness was ~200 μm. GPC and collection conditions were as below:

GPC conditions	
Column	Shodex, KF802.5, 300 mm X 8.0 mm
Mobile phase	THF, 1ml/min
Sample	1 : 1 PEG 1500 / PEG 4600, 0.05%(w/v) in THF
Injection volume	10 μl
Collection conditions	
Nozzle Sheath Gas pressure	30 psi
Nozzle Sheath Gas temperature	208°
Stage velocity	4.6 mm/min

The DHB foil was adhered to a collection plate and placed in the sample compartment of a Perseptive Biosystems Voyager Elite, operated in the reflectron mode. A number of spectra were collected at discrete locations along the deposition track. Figure 5 shows a GPC infrared chromatogram of the sample.

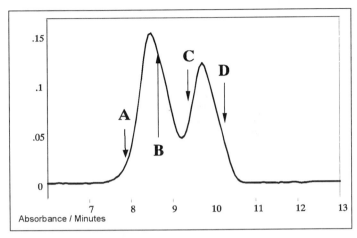

Figure 5 (Infrared) chromatograph of the PEG blend sample.

Two peaks are evident, and are approximately equal in size. Figure 6 is a composite of the mass spectra collected, arranged along the chromatograph time axis.

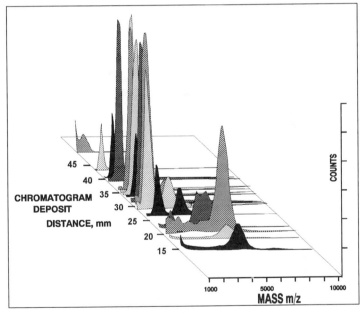

Figure 6 PEG sample spectra of the GPC separation deposit.

Casual inspection of this figure reveals additional complexity within this two component blend. Four spectra, corresponding to the elution times indicated on Figure 5 are shown in Figure 7.

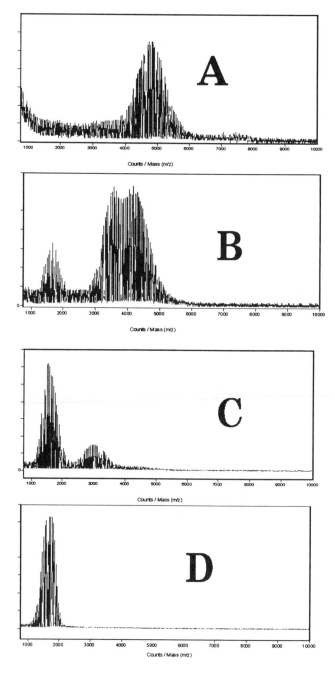

Figure 7 Spectra taken at indicated positions on Figure 5

The earliest eluting component is predominantly material having a centroid at ~4800 mw, but there is also evidence of a small population between 7000 - 8000 mw.

The spectrum at point B shows evidence of three populations, while the spectrum collected at point C shows two populations corresponding to those seen at point B.Populations are monomodal, indicating little fragmentation (a common characteristic of MALDI ionization). The spectrum at point D is clearly monodisperse. The two materials, although polydisperse show little population overlap. These results indicate that the PEG 4600 is slightly dimerized, while the PEG 1500 has a large dimer component. These dimers are approximately double the molecular weight of the respective components. The appearance of these dimers along the time (i.e. mw) axis of the chromatogram is consistent with GPC performance, and is not an artifact of MALDI-MS. Evidence of the dimers by examination of a conventional chromatogram is scanty and uncertain, but GPC - MALDI clearly reveals the true composition of this blend sample.

Figure 8 is a composite of a similar experiment wherein a PMMA polymer standard was examined by GPC - MALDI.

This material is nominally a 3100 mw narrow dispersity standard. The MALDI data indicates a rather broader dispersity than would be expected, and the evidence of a lower weight oligomer in the late eluting portion of the peak.

Discussion

The combination of GPC and MALDI-MS clearly provides the polymer analyst with a powerful new tool for polymer characterization. In the low molecular weight domain, MALDI is capable of much higher resolution than can be obtained by classical GPC methods, hence much more detailed insights into polymer structure and formation. Although MALDI-MS can provide spectra of high molecular weight species the ability to resolve individual oligomers diminishes with increasing molecular weight. Although detection sensitivity drops with increasing molecular weight, sample quantities are seldom a limitation with synthetic polymers. Heretofore, the bias in sensitivity (as a function of polymer mw) has limited MALDI-MS to samples of rather narrow polydispersity. The coupling of GPC and MALDI-MS eliminates this restriction, because the dispersity of the instantaneous eluant from a GPC column is low. The use of precoated matrix targets and a spray deposition system reduces many of the sample preparation problems associated with MALDI-MS, and provides a rapid automated tool for polymer characterization.

94

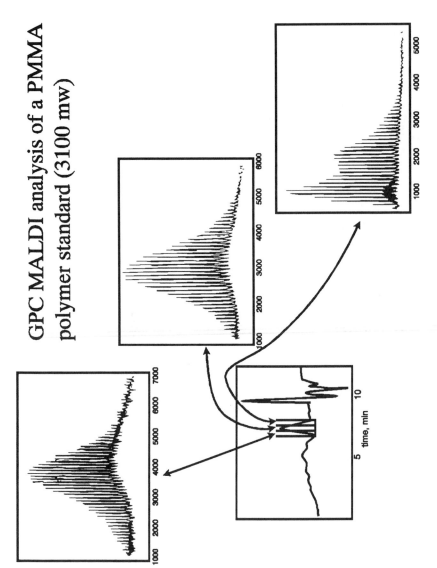

Figure 8 GPC chromatogram of PMMA standard and spectra of three fractions.

Chapter 8

Quantitation in the Analysis of Oligomers by HPLC with Evaporative Light Scattering Detector

Bernd Trathnigg[1], Manfred Kollroser[1], Dušan Berek[2], Son Hoai Nguyen[2], and David Hunkeler[3]

[1]Institute of Organic Chemistry, Karl-Franzens-University at Graz, A-8010 Graz, Heinrichstrasse 28, Austria
[2]Polymer Institute, Slovak Academy of Sciences, Bratislava, Slovakia
[3]Department of Chemistry, Swiss Federal Institute of Technology, Lausanne, Switzerland

The performance of the evaporative light scattering detector (ELSD) in the analysis of polyethers is critically evaluated with respect to quantitative reliability. A comparison of three different types of instruments with a different design shows, that all of them do not fulfil the expectations of chromatographers and the promises of producers: their response depends quite strongly on mobile phase composition. This problem can be overcome by combination of the ELSD with a density detector.

The goals of any chromatographic analysis are a good separation of the components of a sample, their identification, and their accurate quantitative determination. While this may be comparatively easy with low molecular compounds, for which a calibration is easily obtained, a quantitatively accurate characterization of polymers and oligomers is complicated by the fact, that a sufficient resolution can be achieved at best for lower oligomers. The separation of higher oligomers typically requires gradient elution.

Depending on the nature of the samples (and the chromatographic technique), different detectors can be applied.

The most familiar detectors are the UV-detector, which can, however, only be applied to samples absorbing light of a wavelength, for which the mobile phase is sufficiently transparent, and the Refractive Index (RI) detector. The density detector (according to the mechanical oscillator principle) is very useful in polymer analysis, especially in combination with other detectors. Both the density and RI detector can only be applied in isocratic elution.

Consequently, the analysis of samples without chromophores - such as aliphatic polyethers - by gradient elution faces a severe detection problem. In the last years, the Evaporative Light Scattering Detector (ELSD) [1-3] has become a very promising tool for such analytical tasks[4-6]. It is claimed to be a „mass detector", because is should detect any non-volatile material in any mobile phase composition.

Unfortunately, this is not true: the sensitivity of this instrument depends on various parameters[7] which can not always be easily controlled, and its response to polymer homologous series is not as well understood as that of RI and density detector.

Moreover, the response of such an instrument is generally not linear with concentration [7, 8] but can be expressed by an exponential relation [2, 9]. In a previous paper[10] we have studied the linear range of ELSD, density and RI detector for hexaethylene glycol. While the response of density and RI detector was perfectly linear with concentration, this is typically not the case with the ELSD. Moreover, the response factors of different oligo(ethylene glycol)s depended in a different way on the flow rate of the mobile phase and the inlet pressure of the carrier gas. This is no wonder, if one considers the complicated way from the eluate to the detector signal.

In such an instrument, the eluate is nebulized by a stream of air, nitrogen or another carrier gas, the nature and flow rate of which may affect the sensitivity. It must be mentioned, that there are basically two different designs: in the SEDEX and DDL 21 instruments, the mobile phase is nebulized at room temperature in a special spray chamber, in which larger droplets are trapped, while in other types (PL, ALLTECH, DDL 31) the entire aerosol thus obtained passes the heated drift tube, where volatile components are (more or less) evaporated. Obviously, the number and size of the droplets (and in the SEDEX instruments also the nebulized fraction of the eluate) depend on the composition and the flow rate of the mobile phase as well as on the flow rate of the carrier gas[2, 11].

The degree of evaporation depends on evaporator temperature as well as on the flow rate of the carrier gas, which determines the time a droplet spends in the evaporator. At the end of the evaporator tube the particles remaining in the gas stream after evaporation of the mobile phase scatter a transversal light beam.

The intensity of the scattered light depends on number and size of the scattering particles, and should reflect the amount of non-volatile material eluted from the column within each section of the chromatogram. As number and size of the particles depend not only on the concentration of a solute in the eluate, but also on other parameters influencing the original size of droplets formed in the nebulizer, it is clear, that the response of an ELSD is affected by the operating parameters chosen by the operator (the pressure of carrier gas, which determines the gas flow, the temperature of the evaporator, and the photomultiplier gain)[2, 7, 10, 12, 13].

In this study, three different instruments were applied to the same analytical task and their behaviour compared.

Due to the different concepts, the operating conditions for the three detectors had to be chosen properly in order to make them comparable.

As has already been shown in a previous paper[10], lower oligomers are lost, if the evaporator temperature is too high, which means in some cases higher than 30°C. In the case of the PL detector, which leads the entire aerosol into the evaporator, larger droplets will, however, not be evaporated completely at such low temperatures. Unfortunately, in the PL instrument temperature can not be selected freely, but is determined by the software.

Moreover, the carrier gas stream can be controlled in different ways (inlet pressure or gas flow), so we have tried to select the optimum parameters for each instruments (in order to have fair conditions) and keep them constant for all measurements.

Experimental:

The measurements were performed using different equipment:

The SEDEX 45 (Sedere, France) was combined with density detection system DDS 70 (Chromtech, Graz, Austria). Nitrogen was used as carrier gas for the ELSD, and the pressure at the nebulizer was set to 2.0 bar for all measurements.

The mobile phase (acetone-water in different ratios, both solvents HPLC grade, from Promochem, Wesel, Germany) was delivered by two JASCO 880 PU pumps (from Japan Spectrosopic Company, Tokyo, Japan), which were coupled in order to provide gradients by high pressure mixing. The flow rate was 0.5 ml/min in gradient and in isocratic measurements. Mobile phases were mixed per weight and degassed in vacuum. In gradient elution, mobile phase A was pure acetone, mobile phase B was acetone-water 80:20 (w/w). The following gradient profile was used: start 100 % A, then in 50 min to 100 % B, 4 min constant at 100 % B, then within 1 min back to 100 % A.

Mobile phase density was determined using a density measuring device DMA 60, equipped with a DMA 602 M measuring cell (both from A.PAAR, Graz, Austria).

The following columns (both from Phase Separations, Deeside, Clwyd, UK)) were used, which were connected to two column selection valves (Rheodyne 7060, from Rheodyne, Cotati, CA, USA):

a)   Spherisorb S3W, 3 μm, 80Å , 150 x 4.6 mm
b)   Spherisorb S5W, 5 μm, 80Å , 250 x 4.6 mm

For bypass measurements, a capillary (500 mm, 0.5 mm inner diameter) was also connected to the valves. Samples were injected using an autosampler Spark SPH 125 Fix (from Spark Holland, Emmen, The Netherlands) equipped with a 50 μl loop.

Data acquisition and processing was performed using the software CHROMA (Chromtech, Graz, Austria).

The PL EMD 960 (from Polymer Laboratories, Church Stretton, Shropshire, UK) and the DDL 21 (from EuroSep, Cergy, France) were used in combination with a Waters Pump 510 (Waters, Milfor, MA, USA). Injected volumes were 50 μl, and a 500 mm capillary with 0.5 mm ID was used in all measurements.

The flow rate was 0.5 ml/min, the temperature was 25.0°C, and the mobile phase was acetone-water 90:10 (w/w) unless mentioned otherwise. Nitrogen was used as carrier gas. Data acquisition and processing was performend with a Waters Maxima/Baseline 810 PC based system.

The individual conditions for the ELSDs were as follows:

PL-EMD 960:

Nitrogen flow rate 5 ml/min, evaporator temperature 50°C, gain 4

DDL 21:

Nitrogen flow rate 0.7 ml/min, evaporator temperature 50°C, gain 500

Polyether samples were purchased from Fluka (Buchs, Switzerland). Monodisperse oligomers (CnEOm) were typically >98% pure. Polydisperse samples were specified by the producer as follows: Brij 30: Polyethylene glycol dodecyl ether, main component: tetraethylene glycol dodecyl ether; Brij 35: Polyethylene glycol dodecyl ether, main component: trikosaethylene glycol dodecyl ether.

Results and discussion:

Figure 1 shows an isocratic separation of Brij 30, a fatty alcohol ethoxylate (FAE), which was obtained on a plain silica column in acetone-water 99:1 (w/w), with density detector and ELSD. Under these conditions, only a few peaks can be separated and integrated. When a gradient is applied, the situation looks much better: of course, only for the ELSD (Figure 2).

With the same gradient profile, even higher oligomers can be separated quite well, as can be seen from Figure 3, which shows a chromatogram of a 1:5 mixture of Brij 30 and Brij 35. The shoulders in this chromatogram are due to a second polymer homologous series in the samples: Both are ethoxylates of a technical 1-dodecanol, which contains also some amount of 1-tetradecanol and traces of 1-hexadecanol, as can be shown by 2-dimensional LC[14].

The first question concerned the temperature of the evaporator: Fig. 4 shows an isocratic chromatogram of MePEG 350, which was obtained in acetone-water 95:5 with density and ELS detection (SEDEX 45) at 30°C. Obviously, even at the lowest temperature, which can be reasonably controlled, the lowest oligomers are strongly underestimated by the ELSD. The situation is considerably better for FAE, as can be seen from the following figures, in which the peak area x_i is plotted versus the mass m_i of sample in the peak.

There seems to be no difference in the response factors between the individual homologous series, and also the linearity seems to be not so bad for the SEDEX 45 (Figure 5) and the PL (Figure 6), only the DDL 21 shows a considerable curvature (Fig.7). In most cases, a 4th order polynomial gave the closest fit.

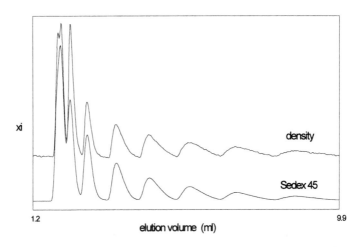

Figure 1: Isocratic separation of Brij 30 (a fatty alcohol ethoxy-late), as obtained on a plain silica column in acetone-water 99:1 (w/w). Detection: density + ELSD

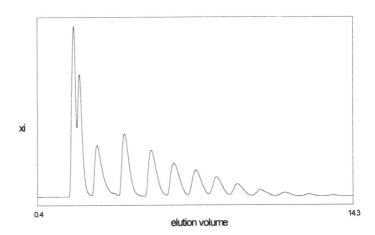

Figure 2: Separation of Brij 30, as obtained on a plain silica col-umn in an acetone-water gradient. Detection: ELSD

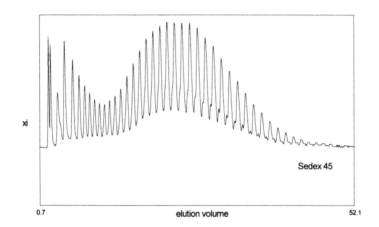

xi

Sedex 45

0.7   elution volume   52.1

Figure 3:   Separation of mixture of Brij 30 and Brij 35 (1:5 w/w), as obtained on a plain silica column in an acetone-water gradient. Detection: ELSD

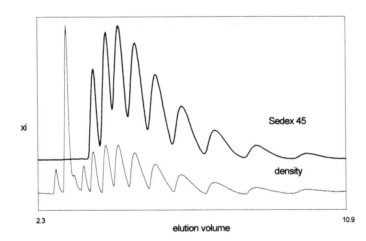

xi

Sedex 45

density

2.3   elution volume   10.9

Figure 4: Isocratic separation of PEG 350 monomethyl ether, as obtained on a plain silica column in acetone-water 99:1 (w/w). Detection: density + ELSD

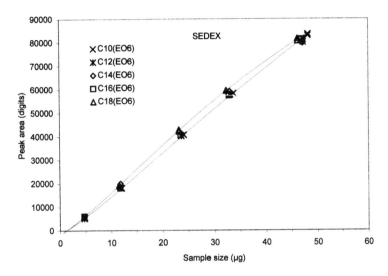

Figure 5: Peak areas (arbitrary units) for hexa(ethylene glycol) monoalkyl ethers, as obtained from bypass measurements in acetone-water 90:10 (w/w) with SEDEX 45

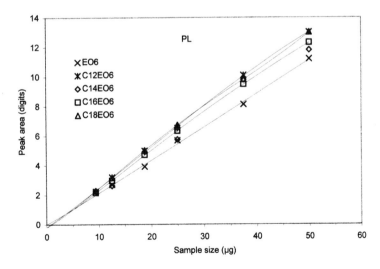

Figure 6: Peak areas (arbitrary units) for hexa(ethylene glycol) and its monoalkyl ethers, as obtained from bypass measurements in acetone-water 90:10 (w/w) with PL-EMD 960

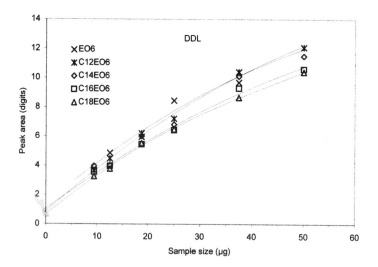

Figure 7:   Peak areas (arbitrary units) for hexa(ethylene glycol) and its monoalkyl ethers, as obtained from bypass measurements in acetone-water 90:10 (w/w) with DDL 21

In the lower concentration range, which is relevant in real separations, an exponential fit can also be applied to describe the relation between peak area x_i and the mass m_i of sample in each peak.

$$x_i = a.m_i^b \qquad \text{Equation 1}$$

The parameters a and b are easily obtained by linear regression in a double logarithmic plot:

$$\ln(x_i) = \ln(a) + b.\ln(m_i) \qquad \text{Equation 2}$$

It must be mentioned, that the exponent b is typically >1 for the SEDEX and <1 for the DDL, while it is very close to 1 for the PL.
The influence of the degree of ethoxylation is also rather small: oligomers with 3 or more oxyethylene (EO) units are quite similar, at least for the SEDEX and the PL, and again the DDL 21 carries the red lantern (Figures 8-10).
The next - and maybe most important - question, however, concerns the influence of the mobile phase composition, which is highly important in gradient elution.
As can be seen in Figures 11-13, the response of all ELSDs in this study depends considerably on the composition of the mobile phase !
This effect is smallest for the PL (maybe due to the higher evaporator temperature!), but it is still too large to be neglected.
For the use of an ELSD in gradient elution, this means, that an accurate quantitation requires the knowledge of the mobile phase composition for each peak,. which is not directly available from the gradient profile.
The determination of mobile phase composition can, however, be easily performed by coupling the ELSD with a density detector.
In Figure 14, the density of acetone-water mixtures at 25.00° C is plotted vs. their composition: over the entire range from 0 to 100 %, a 4th order polynomial describes the relation very well.
In the composition range covered by the gradient used in this study, a linear fit can be applied (Fig. 15).
The signal of the density detector (the T-value) reflects the density (d) of the mobile phase

$$d = A.T^2 - B \qquad \text{Equation 3}$$

(wherein A and B are constants for each measuring cell), hence it can be used to determine its composition with high accuracy.
Fig. 16 shows a gradient chromatogram of Brij 35 with coupled density and ELS detection. As can be seen, the density change due to the gradient is very much stronger than that caused by the sample peaks, hence the mobile phase composition for each peak can easily be obtained from the signal of the density detector. Moreover, the performance of the gradient pump can also be evaluated.

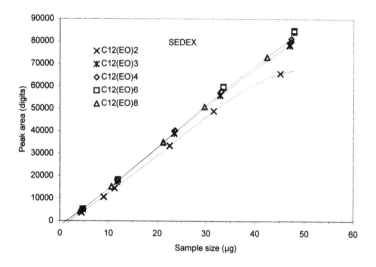

Figure 8: Peak areas (arbitrary units) for oligo(ethylene glycol) monododecyl ethers, as obtained from bypass measurements in acetone-water 90:10 (w/w) with SEDEX 45

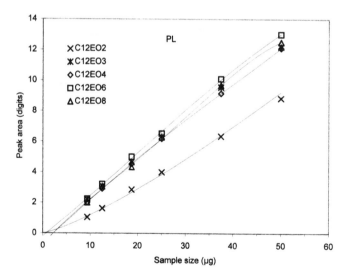

Figure 9: Peak areas (arbitrary units) for oligo(ethylene glycol) monododecyl ethers, as obtained from bypass measurements in acetone-water 90:10 (w/w) with PL-EMD 960

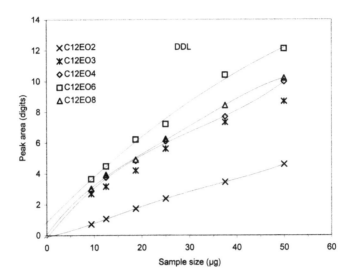

Figure 10: Peak areas (arbitrary units) for oligo(ethylene glycol) monododecyl ethers, as obtained from bypass measurements in acetone-water 90:10 (w/w) with DDL 21

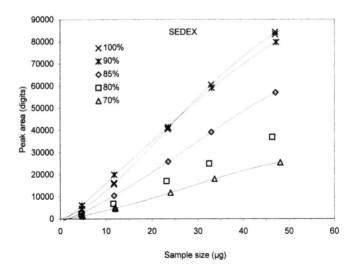

Figure 11: Peak areas (arbitrary units) for hexa(ethylene glycol) mono-tetradecyl ether, as obtained from bypass measurements in acetone-water mixtures of different composition with SEDEX 45

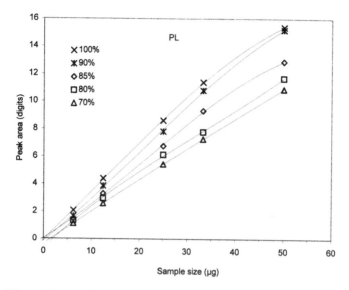

Figure 12: Peak areas (arbitrary units) for hexa(ethylene glycol) mono-tetradecyl ether, as obtained from bypass measurements in acetone-water mixtures of different composition with PL-EMD 960

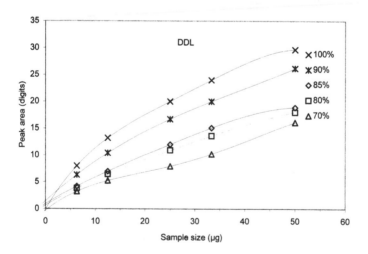

Figure 13: Peak areas (arbitrary units) for hexa(ethylene glycol) mono-tetradecyl ether, as obtained from bypass measurements in acetone-water mixtures of different composition with DDL 21

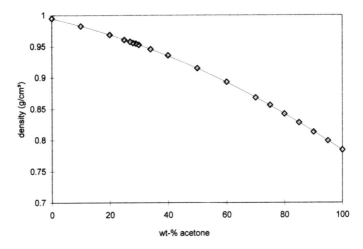

Figure 14:   Density of acetone - water mixtures as a function of
composition in the entire range from 0 to 100 %

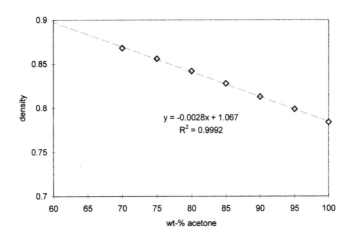

Figure 15:   Density of acetone - water mixtures as a function of
mobile phase composition in the range covered by the gradient

108

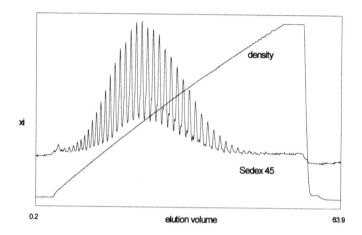

Figure 16: Chromatogram of Brij 35, as obtained by gradient elution with density and ELS detection (SEDEX 45).

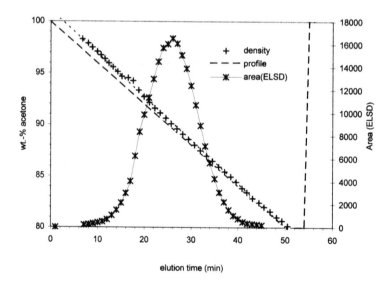

Figure 17: Results from gradient elution in Fig.16: peak areas of ELSD (SEDEX 45) and composition of the mobile phase (as obtained from density detection) for all peaks, compared to the expected mobile phase composition (from the gradient profile)

In Figure 17, the peak areas from ELSD and the mobile phase compositions (from density detection) for each peak are plotted together with the composition calculated from the gradient profile. Obviously, the lines describing the composition do not coincide.

In order to determine the oligomer distribution in such a sample accurately, one would have to determine individual calibrations for all oligomers in different mobile phase compositions, which is not only laborious, but in most cases impossible because of lacking monodisperse oligomer samples.

In practice, only a sufficiently good approximation can be achieved. For this purpose, several assumptions must be made, which appear to be justified by the results shown in Figures 5-10:

1. The response of an ELSD for FAE can be considered to be independent of the end group.
2. The response of ethoxylates with more than 2 EO units is constant.
3. The relation between peak area and concentration in a peak can be described by equation 1 with the same parameters in a given mobile phase composition.
4. The slopes and intercepts in a plot of $\ln(x_i)$ vs $\ln(m_i)$, b and $\ln(a)$ show the same dependence on the composition of the mobile phase for all oligomers.

As can be seen from Fig. 18, straight lines are obtained for the individual mobile phase compositions. In Figure 19, the slopes and intercepts of these lines are plotted versus the composition of the mobile phase. A reasonable linear dependence is found in this plot.

From slope and intercept thus obtained, the parameters a and b in Equations 1 and 2 were calculated for each mobile phase composition.

Using these parameters, the masses of the individual oligomers in each peak of the chromatogram shown in Figure 16 were calculated from the corresponding peak areas.

The effect of such a compensation for mobile phase composition becomes obvious from Figure 20, in which the weight fractions of the individual oligomers (with and without correction) are plotted versus the elution volume.

Conclusions:

The ELSD is a useful instrument in HPLC of polymers, because it allows gradient elution. Very careful calibration work is, however, the precondition for obtaining quantitatively reliable data. The performance of different designs is considerably different. Anyway, all instruments show a considerable dependence of detector response on mobile phase composition. The accuracy of the results can be improved by combination of the ELSD with a density detector, from which the composition of the mobile phase is obtained for each peak.

Acknowledgement

Financial support by the Austrian Academy of Sciences (Grant OWP-53) is gratefully acknowledged.

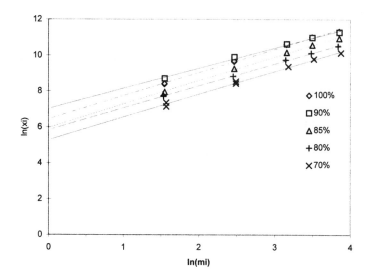

Figure 18: ln-ln-plot of peak area x_i and sample size m_i (in μg) of C14(EO)6 (equation 2), as obtained from bypass measurements (with SEDEX 45) in different mobile phase compositions

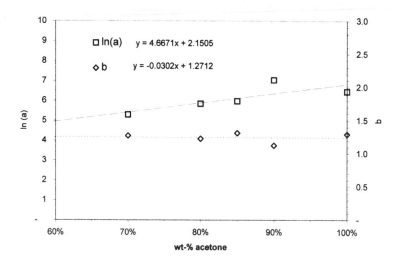

Figure 19: Slope (ln a) and intercept b (equation 2) from Fig.18 as a function of mobile phase composition

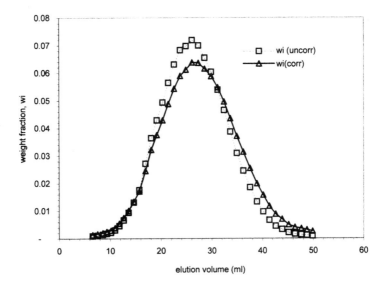

Figure 20: Weight fraction w_i of the individual peaks in gradient LC of Brij 35 with ELSD SEDEX 45, with and without correction of response factors for mobile phase composition

112

References:

1.) Dreux, M. and Lafosse, M.; *Analusis* **1992,** *20,* 587-595.
2.) Dreux, M., Lafosse, M. and Morin-Allory, L.; *LC-GC Internatl.* **1996,** *9,* 148-156.
3.) Lafosse, M., Elfakir, C., Morin-Allory, L. and Dreux, M.; *Hrc J High Res Chromatogr* **1992,** *15,* 312-318.
4.) Rissler, K., Fuchslueger, U. and Grether, H. J.; *J Liq Chromatogr* **1994,** *17,* 3109-3132.
5.) Desbene, P. L. and Desmazieres, B.; *J Chromatogr a* **1994,** *661,* 207-213.
6.) Brossard, S., Lafosse, M. and Dreux, M.; *J Chromatogr* **1992,** *591,* 149-157.
7.) Van der Meeren, P., Van der Deelen, J. and Baert, L.; *Anal Chem* **1992,** *64,* 1056-1062.
8.) Hopia, A. I. and Ollilainen, V. M.; *J Liq Chromatogr* **1993,** *16,* 2469-2482.
9.) Koropchak, J. A., Heenan, C. L. and Allen, L. B.; *J Chromatogr a* **1996,** *736,* 11-19.
10.) Trathnigg, B., Kollroser, M., Berek, D. and Janco, M.; *Abstr Pap Amer Chem Soc* **1997,** *214,* 220-PMSE.
11.) Van der Meeren, P., Van der Deelen, J., Huyghebaert, G. and Baert, L.; *Chromatographia* **1992,** *34,* 557-562.
12.) Mengerink, Y., Deman, H. C. J. and Van der Wal, S.; *J Chromatogr* **1991,** *552,* 593-604.
13.) Miszkiewicz, W. and Szymanowski, J.; *J Liq Chromatogr Relat Techno* **1996,** *19,* 1013-1032.
14.) Trathnigg, B. and Kollroser, M.; *Intern.J.Polym.Anal.Char.* **1995,** *1,* 301-313.

FIELD FLOW FRACTIONATION AND COUPLED LIQUID CHROMATOGRAPHY METHODS

Chapter 9

Advantages of Determining the Molar Mass Distributions of Water-Soluble Polymers and Polyelectrolytes with FFFF–MALLS and SEC–MALLS

W.-M. Kulicke, S. Lange, and D. Heins

Institut für Technische und Makromolekulare Chemie, Universität Hamburg, Bundesstrasse 45, D-20146 Hamburg, Germany

This paper describes the characterization of the absolute molar mass distribution of water-soluble polymers with the combined fractionation apparatus of size-exclusion chromatography (SEC) and flow field-flow-fractionation (FFFF) coupled with a multi-angle laser light-scattering (MALLS) photometer, which is sensitive to molar mass, and a differential refractometer to measure the differential refractive index (DRI), which is sensitive to concentration. Emphasis is placed on the advantages of these methods of determination with reference to polymers having a variety of structures, such as polysaccharides, polycations, polyanions and synthetic polymers. It is also shown how the separate fractions can be characterized during enzymatic degradation. The same is true for ultrasonic degradation. In the examples illustrated here degradation occurs in the centre of the chain so that homologous series are generated while at the same time the molar mass distribution becomes somewhat narrower. The cause of this is the asymmetric molar mass distribution of the native sample. Examples are used to discuss the merits and limitations.

On account of their properties, water-soluble synthetic and biological polymers and polyelectrolytes have commercial applications in a large number of technological fields, examples include use as flow enhancers, thickening agents and stabilizers (1). Both synthetic polymers and those based on renewable raw materials are mixtures of homologous substances with differing molar masses. For the former this is a consequence of the statistics inherent in every polymer reaction, for the latter the reason is to be found in the lack of reproducibility within nature, and in degradation reactions during pulping or derivatization. The molar mass of a polymer is the product of the molar mass of the monomer, M_{mon}, (basic component/repeating unit) and the degree of polymerization, P, i.e. the number of repeating units in the respective polymer chain. For relatively low degrees of polymerization (P < 100) the molar mass distribution can still be described and detected analytically as a distribution of individual species, whereas the molar mass distribution for higher degrees of polymerization turns into a quasi-continuous, generally asymmetric distribution function (see Figure 1). The distribution of molar mass and particle size has a crucial influence on the property profiles of polymers in solution. This paper aims to take an in-depth look at the often problematic determination of the distributions.

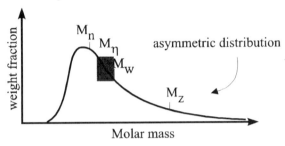

Figure 1. Schematic representation of an asymmetric distribution function.

The determination of molar mass has been a very time-consuming and expensive process, so that mean values are generally given to characterize polymers. These mean values (Figure 1) depend upon the method of determination, hence for example the number-average, M_n, is determined from osmometric measurements, the weight-average, M_w, from light-scattering measurements and the viscosity average, M_η, from viscosity measurements. It is known that the mean of M_η (grey shading in Figure 1) varies as a function of the exponent a of the Mark-Houwink relationship (2). The ratio of M_w to M_n (M_w / M_n) is often given as a measure of the polydispersity.

In some special applications the declaration of mean values is not sufficient for product optimization and quality assurance. Only knowledge of the entire distribution curve can lead to a distinct characterization of the products and thus allow the structure and technological properties to be correlated. For instance, the length of the soft segments, i.e. diol component, has a crucial influence on the macroscopic properties of segmented polyurethanes (3). The broad distribution of polyethylene is known to be decisive for the quality of processing (Ref. 2, P. 246). Knowledge of the overall distribution assumes almost vital importance for plasma substitutes as high-molar-mass flanks are suspected of triggering anaphylactoid reactions, and mean values do not yield the crucial information (4).

One way of determining the molar mass distribution is to fractionate the sample and then determine the molar mass of each separate fraction.

Size exclusion chromatography (SEC) (5,6,7) and recently Flow Field-Flow Fractionation (FFFF) (8,9,10) too, can be employed to fractionate polymers by their size. Coupling these fractionating units with a detector system consisting of a light-scattering photometer (MALLS) and a differential refractometer (DRI) makes it possible to separate the polymers and at the same time to carry out an absolute determination of the molar mass and radius of gyration, (R_G); hence the entire distributions are determined for molar mass and radius (4,11).

116

This paper aims to demonstrate that these two methods (SEC/MALLS/DRI and FFFF/MALLS/DRI) can be used to determine the distributions of molar mass and radius of gyration for many water-soluble polymers.

Polymers investigated

Bovine serum albumin (BSA, Fluka, Neu-Ulm, Germany) is a member of the large albumins group. Together with the globulins and prolamines, these form the most important group of proteins. BSA is obtained from the blood of cattle and has a molar mass of about 66,000 g/mol (*12*). The refractive index has been determined as 0.170 mL/g (Wood RF-600 (Wood Co., PA, USA), 0.1 M NaNO$_3$ solution with 0.02% azide added, 633 nm, 298 K).

The **tobacco mosaic virus** (TMV) consists of a ribonucleic acid helix which is stabilized by 2,130 protein sub-units that are suspended in an outward direction. The result is a hollow cylinder with a length of 300 nm and a width of 15-18 nm, the inside diameter is approx. 4 nm (*13*). The TMV investigated here was kindly provided by M. Mackay, University of Queensland, Australia. A transmission electron micrograph of the tobacco mosaic virus is shown in Ref. *14*.

The **starch derivatives hydroxyethyl starch and acetyl starch** were prepared from so-called waxy starches with an amylopectin content of > 95 % (*15,16*). Amylopectin is the highly branched component of starch and its aqueous solutions are stable. The main chain consists of α-(1→4)-linked D-glucose, with α-(1→6)-linked branching site every 18 to 27 glucose residues (see Figure 2) (*17*). The desired mean molar mass is adjusted by partial hydrolysis.

The hydroxyethyl starch samples investigated were commercial products used as plasma substitutes. Hydroxyethylation was carried out by means of ethylene oxide in alkaline medium (*18*). Esterification to acetyl starch was accomplished by means of acetic anhydride in alkaline medium (*16*).

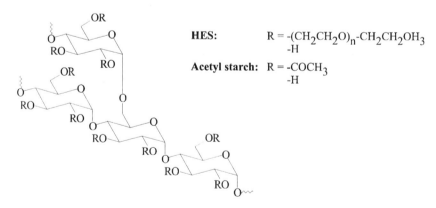

Figure 2.   Structure of the starch derivatives acetyl and hydroxyethyl starch: α-(1→4) - linked chain of glucose units with α-(1→6) - branches.

The refractive index increments were determined in 0.1 M NaNO$_3$ solution with 0.02% azide added (T = 398 K) at 0.133 mL/g for hydroxyethyl starch and at 0.138 mL/g for acetyl starch.

The **cellulose derivatives hydroxyethyl cellulose (HEC) and carboxymethyl cellulose (CMC)** (Figure 3) were synthesized from alkaline cellulose as the starting material. This involved activating the cellulose with caustic soda in an inert solvent (e.g. isopropanol), i.e. the hydrogen bonds in cellulose are stretched or broken (*19*).

Hydroxyethylation to produce HEC was performed by epoxide reaction with ethylene oxide at a temperature of 30-80°C. The maximum average degree of substitution for cellulose is DS = 3.0 because the substituent can also react with the reagent the molar degree of substitution (MS) may be higher. For commercial samples the DS lies within a range from 0.8 to 1.2, whereas the MS takes values from 1.7 to 3.0. An MS of 2.5 was given for the samples from Polysciences investigated here. The refractive index increment was determined as 0.145 mL/g.

Carboxymethylation to CMC was carried out by a Williamson ether synthesis in a heterogeneous reaction using chloroacetic acid and involving the formation of NaCl. The samples investigated here came from the company Wolff Walsrode AG. Sample CMC 1 has a DS of 1, sample CMC 2 a DS of 2.4. A refractive index increment of 0.136 mL/g was used.

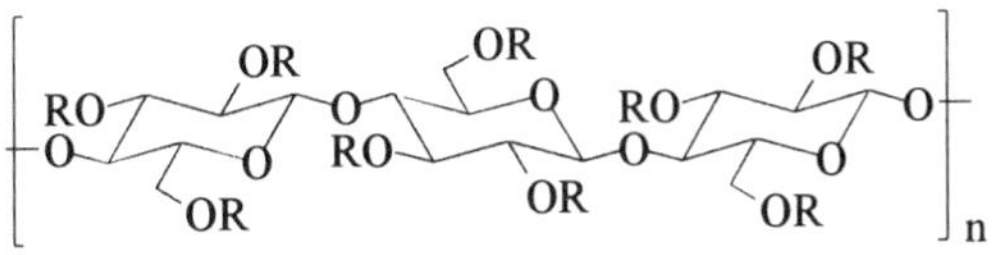

HEC: R = -(CH$_2$CH$_2$O)$_n$-CH$_2$CH$_2$OH$_3$ CMC: R = -CH$_2$COO$^-$ Na$^+$
      -H   -H

Figure 3. Structure of the cellulose derivatives: β-(1→4) - linked chain of glucose units.

The **sodium polystyrene sulphonate standards (NaPSS)** (left-hand formula in Figure 4) used come from the company Polymer Standard Service (Mainz). According to the distributor, the samples were manufactured by anionic polymerization of styrene followed by sulphonation. The degree of sulphonation is given as greater than 90% and the polydispersity as M_w / M_n less than 1.1 (M_w / M_n (NaPSS-7) < 1.3). According to the information supplied, the samples underwent dialysis and freeze-drying prior to delivery. The refractive index increment was determined after Equilibrium dialysis in 0.1 M NaNO$_3$ solution with 0.02% azide added (633 nm, 298 K) as 0.195 mL/g (*20*).

The non-ionic **polyacrylamide (PAAm)** (right-hand formula in Figure 4) was synthesized in the laboratory by the radical polymerization of acrylamide, for details refer to Kulicke in Houben-Weyl amongst others (*21,22*). A value of 0.177 mL/g was determined for the refractive index increment after Equilibrium dialysis in 0.1 M NaNO$_3$ solution with 0.02% azide added (633 nm, 298 K).

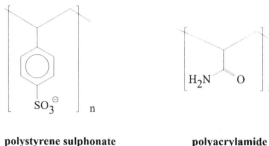

polystyrene sulphonate   **polyacrylamide**

Figure 4.   Monomer units of polystyrene sulphonate (left) and polyacrylamide (right).

Poly(diallyldimethylammonium chloride) (poly-DADMAC) is a cationic polymer that is used in water treatment (flocculation and dewatering).

Synthesis is by radical cyclopolymerization in aqueous solution (*23*). The product contains the three possible structures shown in Figure 5. The refractive index increment was determined after Equilibrium dialysis in 0.1 M $NaNO_3$ solution with 0.02% azide added (633 nm, 298 K) as 0.12 mL/g (*16*).

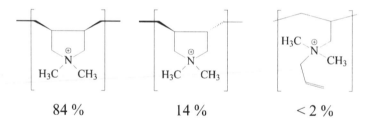

84 %   14 %   < 2 %

Figure 5.   Structure of the three monomeric units that occur in poly-DADMACs.

Experimental

Sample preparation. In water-soluble synthetic and biological macromolecules there are often undissolved components and gel structures which make characterization more difficult. One of the causes for this behaviour, which is familiar from the applications of technical chemistry, biology and medicine, is that association and aggregation occur due to the dipole-dipole interactions and / or hydrogen bonds. This complicates sample characterization. Pretreatment such as filtration, centrifugation, the use of guard columns, etc. can lead to distortions to such an extent that the samples are no longer fully investigated, but only partially. To enable the results for different samples to be compared, the method of sample preparation has been described in detail.

Wherever possible, the same solution was used for the solvent as for the eluent. The water used was deionized and then distilled. With the exception of the tobacco mosaic virus, which was analysed in 0.04 M sodium dodecyl sulphate solution, all the other samples were analysed in 0.1 M sodium nitrate solution. All the solvents were treated with 200 ppm sodium azide as a bactericide and also underwent pressure filtration using a filter of 0.1 μm pore size.

The samples analysed by SEC/MALLS underwent ultracentrifugation (1 h, 30,000 g) and in-line filtration (pore size: 0.8 μm) to remove any coarse impurities. This pretreatment was not necessary for the readings conducted with FFFF/MALLS.

Ultrasonic degradation. Homologous series of polymers are required to establish structure-property relationships ([η] - M_W - relationships). These can be produced by selective degradation of a molecule. However, this presupposes that side reactions such as the elimination of side-groups can be avoided. Degradation in a ball mill, by means of high shear forces or the effect of temperature does not fulfil this condition. Kulicke et al. have shown that elongation flow processes, as occur in ultrasonic treatment, lead to reproducible chains scission without any undesired side reactions (*24*).

Reduction of the molar mass by means of ultrasound is based on the fact that the sound waves cause pressure fluctuations in the solution. Local pressure differences vaporise the solvent thus generating gas bubbles in the solution, which grow in size during further stages of compression and expansion, before collapsing beyond a certain size. This process, termed cavitation, generates a flow field in which very high shear rates occur (*25*). The stretching takes place uniformly, and on being stretched to approx. 20 %, the polymer chain is broken close to the centre of the macromolecule, where the force exerted is at its greatest (*26*).

Ultrasonic reduction of the molar mass was performed with a W-450 Branson Sonifier (Branson Schallkraft GmbH, Heusenstamm, Germany) using a 19-mm titanium resonator at a frequency of 30 kHz on the medium power setting. In each case 13 mL of an approx. 0.2 % sample of the solution were treated. The solution was conditioned to ~ 293 K. The particles abraded from the resonator were removed by centrifugation (1 h / 30,000 g).

Enzymatic degradation. Plasma substitutes are adjusted so that they match the desired properties exactly, such as volume effect, which is the volume expansion due to the influx of the tissue fluid caused by the colloid-osmotic pressure, or the retention time, which is the time the substance stays in the bloodstream before being excreted. The volume effect depends upon the concentration or the number of molecules, the retention time upon the molar mass, the degree of substitution and the kinetics of degradation. In order to investigate the last of these factors, an acetyl starch and a hydroxyethyl starch with identical nominal masses and degrees of substitution were investigated with regard to enzymatic degradation induced by α-amylase.

Hydroxyethyl starch is employed medically in 3 %, 6 % and 10 % solutions. Solutions containing 3 % of the two starch derivatives were made up in double distilled water. These were added to the temperature-conditioned haematoid buffer solution (phosphate buffer, pH = 7.4, T = 310 K, α-amylase activity: 144 U/L). After a defined time interval the degradation reaction was interrupted by raising the temperature to 358 K, at which temperature the α-amylase is irreversibly deactivated.

Fractionating methods

In many problems it is not enough to simply give the mean molar masses, their distribution is also relevant. One established method for determining the molar mass distribution is that of size exclusion chromatography (SEC). However, in some applications this technique meets its limitations. Flow field-flow fractionation (FFFF) is an alternative method for determining molar mass. Separation with SEC functions on the basis of the hydrodynamic size of the polymers, whereas FFFF is based on the different diffusion coefficients. The following section will describe the two methods.

Size exclusion chromatography (SEC). SEC is a special form of HPLC (high performance liquid chromatography) in which ideally no interactions with the stationary phase occur and in which the sample is separated within the column according to its hydrodynamic volume only. Porous glass beads or cross-linked gels with a defined pore size serve as the separating medium. During separation the large particles are eluted first. These are less able to penetrate into the pores, they thus have a lower volume at their disposal and their effective free path length becomes shorter. Particles that are too large to penetrate into even the largest pores are excluded due to size and hence eluted before the actual measurement begins; they are therefore not included in the separation process. Small particles are able to penetrate further into the pores and thus have more volume at their disposal. As the diffusion process is the only influential factor inside the pores, the elution of small particles is delayed. A particle inside a pore is only able to get back into the eluent stream by diffusion.

The concentration of each eluted fraction is determined. With suitable calibration the position of the peak in the elution graph can now be used for relative determination of the molar mass. However, it is necessary here to assume that identical solution states, i.e. the same hydrodynamic radii, apply to corresponding molar masses. The most convenient method of calibration uses the same polymer/solvent system. This presents a problem since there are many polymers for which no narrowly distributed standards are available. Universal calibration based on the connection between hydrodynamic volume, Staudinger index and molar mass often fails due to the lack of a reliable Mark-Houwink relationship. Figure 6 provides a schematic illustration of two differently sized particles being separated and the procedure in determining the relative molar masses by means of SEC.

In practice problems may occur when SEC is used for separation. Apart from the size exclusion already mentioned, various forms of interaction with the stationary phase may take place. Unlike charges on polymer and stationary phase may result in partial adsorption (Figure 7 A), the polymers are then eluted later or may remain stuck in the column. In contrast, like charges on separating medium and polymer may lead to ion exclusion, i.e. the pore becomes effectively smaller and the particles are eluted earlier than their hydrodynamic radius would lead one to expect (Figure 7 B). It is particularly in relative SEC that these interaction lead to erroneous results, which can be recognized by absolute determination. Some of these effects may be suppressed by adding an electrolyte. A further problem is that depending on the flow selected, degradation may occur because of elongation flow currents (Figure 7 C) *(27)*.

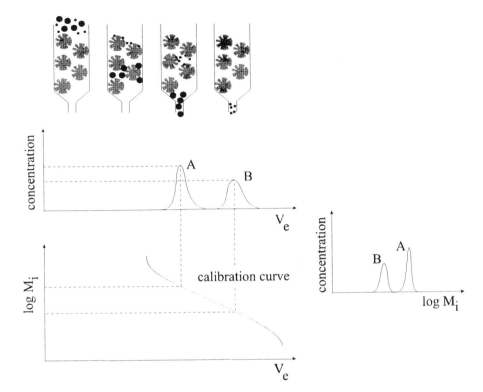

Figure 6. Schematic representation of the separating mechanism in SEC and of the relative molar mass determination for two particles with differing hydrodynamic radii.

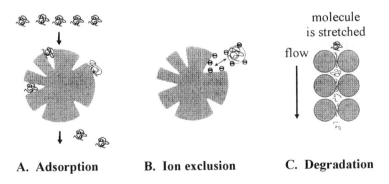

A. Adsorption   **B. Ion exclusion**   **C. Degradation**

Figure 7. Schematic representation of disruptions to the ideal SEC separating mechanism.

Separation of the polymers by size was carried out over four TSK PW$_{XL}$ columns from the company TosoHaas (Stuttgart), which were arranged in order of decreasing pore size. To protect the columns, the samples first passed through a guard column of the same material.

Flow Field-Flow Fractionation (FFFF). The separating principle of FFFF (8-10) differs fundamentally from that of SEC. In the cross flow channel the sample constituents are separated in a hydrodynamic force-field according to their diffusion coefficients (Figure 8). During a relaxation phase with the channel flow switched off (stop-flow-relaxation) the molecules occupy an equilibrium position corresponding to their diffusion coefficient. After completion of this process, the molecules are eluted at different speeds according to their parabolic flow profile. The strength of the field of forces – and hence the separating power – may be varied within wide limits beyond the cross flow and may for instance even be altered *during* the measurement (Programmed Field of Force / PFF).

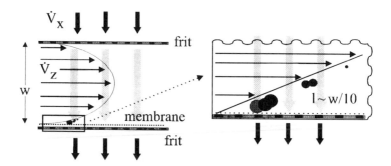

Figure 8. Schematic longitudinal section through the cross flow separating channel. Three samples are sketched in that are eluted at different speeds according to the different diffusion coefficients of the parabolic flow profile.

The higher the polydispersity of the sample, the more accurately the cross flow must be adjusted to the sample, in order to detect both the smallest and also the largest constituents. In addition, the diffusion coefficients can also be calculated from the retention time, t_r, via a theoretical relationship (28).

$$D = \frac{w^2}{6 \cdot t_r} \cdot \frac{\dot{V}_x}{\dot{V}_z} \tag{1}$$

where: $\dot{V}_z$ channel flow   w   channel height
        $\dot{V}_x$ cross flow t_r retention time

The phenomenon of size exclusion as in SEC does not occur here. However, beyond a size of approx. 1 μm a change in the elution mode takes place and the particles are no longer separated by diffusion coefficient (normal elution), but by their size (steric elution) (*29*). The lower limit of the molecules to be separated is given by the membrane lying on the lower frit (here exclusion limit 10,000 g/mol for dextran).

Pre-cleaning of the polymer solutions is not necessary as coarse impurities are eluted before the sample peak in the so-called void peak and are thus taken out of the sample.

These advantages should make FFFF a suitable separating method for cellulose derivatives in particular, as some of these have a high polydispersity and in addition to components dissolved in a molecularly disperse form also contain associations and aggregates.

Absolute molar mass determination by means of light-scattering (MALLS)

Light scattering is based on the interactions between matter and electromagnetic radiation. Electromagnetic waves induce oscillations in electrons. These oscillations have the same frequency as the primary radiation, and the consequence is light scattering. For particles with a diameter of less than 1/20 of the incident wavelength, λ_0, the intensity of the light scattering is independent of the observation angle, in larger particles interference phenomena mean that angular dependency is seen. Assuming that the polymer concentration in the light-scattering cell is small, the following relationship may be established between the reduced scattered light intensity, R_9, and the weight-average molar mass, M_W:

$$\frac{K \cdot c}{R_9} = \frac{1}{P(9)} \cdot \left(\frac{1}{M_W} + 2 \cdot A_2 \cdot c + ... \right) \tag{2}$$

Where
$$\frac{1}{P(9)} = 1 + \frac{1}{3} \cdot q^2 \cdot R_G^2 \tag{3}$$

And
$$q = \frac{4 \cdot \pi}{\lambda_0} \cdot \sin\left(\frac{9}{2}\right) \tag{4}$$

where K, the so-called light-scattering constant, contains the wavelength of the primary radiation, λ_0, the refractive index of the pure solvent, n_0, and the refractive index increment, dn/dc; c is the concentration. $P(9)$ takes into account the angular dependency of the scattered light intensity. Plotting $(K \cdot c)/R_9$ against $\sin^2(9/2)$ and taking the reciprocal of the intercept of the axes leads to the weight-average molar mass.

The light scattering measurements were performed with a DAWN-F light-scattering photometer from the company Wyatt Technology Corp. (Santa Barbara, USA). An He-Ne laser ($\lambda = 632.8$ nm) served as the light source. The scattered light was measured simultaneously by fifteen photodiodes arranged in the angular range from 26.56° to 144.46° in stationary positions around the cell. Measurement was carried out in a flow-through cell made of highly refractive glass (K5). The equipment parameters were determined by calibration of the 90° angle with pure toluene followed by normalization of the other detectors with gold particles. The programs *ASTRA 2.0* and *EASI* from Wyatt Technology Corp. (Santa Barbara, USA) were used for evaluation of the results.

Determination of concentration by means of differential refractometry (DRI)

A differential refractometer measures refractive index differences between two solutions, generally solution and solvent. Figure 9 shows the schematic set-up of a differential refractometer. For very small differences in the refractive index in the measuring cell and in the reference cell the deflection δ of the measuring beam, 2, in relation to the reference beam, 1, is proportional to Δn.

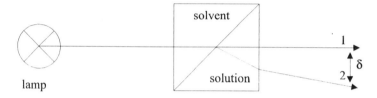

Figure 9. Schematic representation of a differential refractometer, the deflection of the measuring beam, 2, compared with the reference beam, 1, is proportional to the change in refractive index.

For very dilute solutions the measured refractive index differences are proportional to the concentrations, c. The proportionality constant is the refractive index increment (dn/dc).

$$n - n_0 = \Delta n = c \cdot \left(\frac{dn}{dc} \right) \qquad (5)$$

The refractive index increment is a function of the wavelength of the incident light, the temperature, the solvent and the pressure. It is entered into the light scattering constant, K, quadratically and therefore has to be determined as accurately as possible so as to minimize the error in the molar mass determination (see Eqn. 2). A series of concentrations is generally taken for the determination. The measured refractive indices are plotted against the concentration, according to Eqn. 5 the gradient obtained is the refractive index increment.

The concentration of each fraction was carried out with a Shodex RI SE-51 differential refractometer from the company Showa Denko (Tokyo, Japan), which was connected behind the light scattering photometer. A concentration series of NaCl solutions was used for calibrating the apparatus.

Determination of the absolute molar mass distribution

Combining a fractionating method with one for the absolute determination of molar mass makes it possible to carry out an absolute determination of the molar mass distribution. Light scattering is particularly useful for this purpose as it enables the scattering intensities to be measured from continuous flow. Hence for each individual fraction it becomes possible to determine the molar mass and the radius of gyration, R_G, absolutely. If the molar mass is to be calculated for each fraction eluted, it is also necessary to determine the concentration at the same time as the determination of the scattered light intensity. This is achieved by coupling the light-scattering photometer with a concentration detector. Differential refractometry has proven to be a suitable method for this. Apart from being a relatively sensitive method of detection, it also has the advantage that the error which occurs during determination of the refractive index increment is not entered into the calculation of the molar mass (Eqn. 2) quadratically but linearly, as the optical constant is given by $K = f(dn/dc)^2$ and the concentration by $c = f(dn/dc)^{-1}$.

Whereas the coupling of SEC with MALLS/DRI may be regarded as established, coupling FFFF with MALLS/DRI caused considerable difficulty, with the result that the methods were not successfully coupled until 1994.

Figure 10 shows the schematic arrangement of the SEC and FFFF/MALLS/DRI apparatus. To determine the molar masses, 100 to 250 μL of the polymer solution were injected via either the automatic sampler or the injection loop. The SEC-measurements were performed at 298°K and at a flow rate of 0.5 mL/min. The experimental details for the FFFF-measurements are given in the next section.

Characterization of polymers and polyelectrolytes

Accurate knowledge of the molar mass distribution is of great importance in many fields of applied technology. The coupling of SEC with a detector systems consisting of a light scattering photometer and a differential refractometer (SEC/MALLS/DRI) has proven its worth in the determination of molar mass distributions. Not the least important reason for the widespread use of SEC is the fact that it is technically easy to handle. The following will also present results from an apparatus consisting of flow field-flow fractionation, a light scattering photometer and a differential refractometer (FFFF/MALLS/DRI), the coupling of which has only been successfully accomplished very recently (*11*).

Characterization by means of SEC/MALLS/DRI. SEC has proven to be a reliable method in the determination of molar masses and their distributions. However, if there is no direct coupling to a light-scattering photometer, calibration graphs have to be prepared in order to perform the absolute determination. This is further complicated by the fact that suitable standards are not always available and universal calibration then has to be carried out. Coupling the separation apparatus with a light-scattering photometer and a differential refractometer enables the molar mass to be determined absolutely for each fraction.

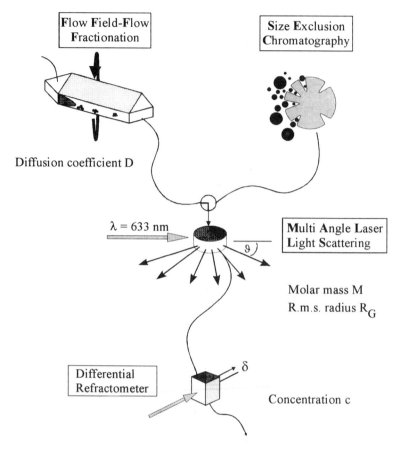

Figure 10. Schematic arrangement of the SEC/FFFF/MALLS/DRI apparatus.

Starch Derivatives. One area where starch derivatives are put to use is in the field of medicine. The starch derivative hydroxyethyl starch (HES) is used as a plasma substitute and has achieved a leading position in Germany with an approx. 90% market share. Their use in medicine requires defined mean molar masses ranging from 40,000 to 450,000 g/mol (*15*). However, it is not only necessary to know the mean molar mass but also the entire distribution. In medicine the sample is required to have a uniform action. If a narrowly distributed and broadly distributed sample are taken, they may have the same mean value but their action will be different. In the sample with the broad molar mass distribution the volume effect (as described above) shortly after the infusion will be very high as the many low-molar-mass particles exert a high osmotic pressure. However, the low-molar-mass constituents are excreted quickly and the volume effect decreases sharply. Furthermore, the especially high-molar-mass constituents are suspected of triggering anaphylactoid reactions.

This was the reason for coupling SEC with light-scattering in order to determine the molar mass distribution accurately. The highly branched compact structure of the amylopectin basic skeleton – as shown by a value of only 0.35 for the exponent of the Mark-Houwink equation – means that high molar-mass components of the sample can be separated by means of SEC. The differential distribution obtained is shown in Figure 11. It can be seen that a sample with a mean molar mass value of $M_W \approx 40,000$ g/mol has components with molar masses as high as 3×10^7 g/mol. This result suggests that the hydroxyethyl starch analysed is unsuitable as a plasma substitute.

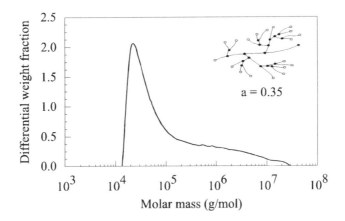

Figure 11. Differential distribution of a HES determined absolutely by means of SEC/MALLS/DRI. The compact structure of the amylopectin basic skeleton (a = 0.35) means that the high-molar-mass component of the sample is also separated.

Acetyl starch is another example of a starch derivative. This too is also being considered for use as a plasma substitute (*30*). The same basic skeleton suggests that similar physiological properties are to be expected, with degradability being increased due to the change in substituents. Knowledge of the molar mass distribution is also essential in this case. Physiological degradability is an important criterion when considering suitability for use as a plasma substitute. In order to investigate this property, an acetyl starch with a nominal molar mass of 200,000 g/mol and a nominal DS of 0.5 was broken down by α-amylase. Haematoid conditions were chosen for the experiment (phosphate buffer, T = 310 K, pH = 7.4).

Figure 12 shows the elution graph of the molar masses and the concentration of the non-degraded acetyl starch. According to the mechanism of size exclusion chromatography, the molecules with the largest hydrodynamic volume, i.e. the highest molar mass, are eluted first and the smaller molecules later. The spread of molar masses in large elution volumes can be attributed to the decrease in scattered light intensity at lower molar masses. The range of detectable molar masses here lies between 4×10^6 and 2×10^4 g/mol. A knowledge of the molar mass and the accompanying concentrations can then be used to determine the molar mass distribution of the sample absolutely, the different means can also be calculated. Figure 13 shows the absolutely determined molar mass distributions of the non-degraded sample and samples that have been degraded for differing lengths of time.

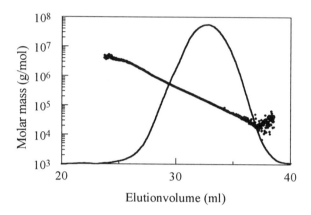

Figure 12. Elution graph of a SEC/MALLS/DRI measurement on the non-degraded acetyl starch. Dependence of the molar mass upon the elution volume superimposed upon the concentration signal (DRI).

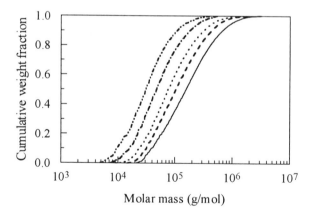

Figure 13. Enzymatic degradation of an acetyl starch by α-amylase. Differential distribution curves determined absolutely by SEC/MALLS/DRI of the non-degraded acetyl starch (--) and after different degradation times (38 min (--), 90 min (⋯), 417 min (-··-), 1367 min(-··-··)).

The accompanying values calculated for the means and polydispersities, M_w/M_n, are given in Table I. Acetyl starch has an initial weight-average molar mass of 2.8×10^5 g/mol. This corresponds to the elution graph in Figure 12. With increasing degradation the distribution curves are displaced to smaller values. After approx. 23 hours the weight-average molar mass has fallen to 5×10^4 g/mol. As expected, the polydispersity falls as the molar mass decreases. The differential distributions in Figure 13 illustrate the fundamental reduction in polydispersity, the polymers become more uniform.

Table I. Compilation of the molar mass and polydispersity values of the natural and the enzymatically degraded acetyl starch samples. The molar mass distributions were measured by SEC/MALLS/DRI.

Degradation time (min)	M_n (g/mol)	M_w (g/mol)	M_z (g/mol)	M_w/M_n (-)
0	1.0×10^5	2.8×10^5	6.8×10^5	2.8
38	7.5×10^4	1.8×10^5	4.1×10^5	2.4
90	6.0×10^4	1.4×10^5	3.2×10^5	2.3
417	3.6×10^4	7.9×10^4	1.7×10^5	2.2
1367	2.5×10^4	5.0×10^4	1.0×10^5	2.0

Carboxymethylcellulose. A further example of a polysaccharide from renewable raw materials is given by cellulose. Investigations into the chemical structure by means of NMR spectroscopy required 10 % solutions. In the case of cellulose and its derivatives these are highly viscous, thus leading to marked broadening of the bands, which makes evaluation impossible (24). In order to make NMR spectroscopic analysis possible in spite of this, attempts are made to lower the solution viscosity by reducing the molar mass of the polymers. One condition for this is that no side-reactions occur.

Both ultrasonic and enzymatic degradation were carried out on a carboxymethylcellulose with a DS of 1 (CMC 1). The molar masses attained were determined with the aid of the SEC/MALLS/DRI system. Figure 14 shows the differential distribution curves. The continuous line represents the native sample while the other lines are for the samples with different degrees of degradation. For the CMC 1 the final molar mass achieved by enzymatic degradation is lower than that obtained with the ultrasonic method. Combining the two methods leads to an even lower molar mass. According to the latest investigations, it is assumed that not only enzymatic but also ultrasonic degradation favours attack at no-substituted regions (31). In contrast to ultrasound, which breaks the polymer near the centre, the enzyme attacks at a position where there are at least three adjacent anhydroglucose units (32). As these occur randomly in the monomer, the same statistics apply to the breaking of the bonds. The result is a broader distribution.

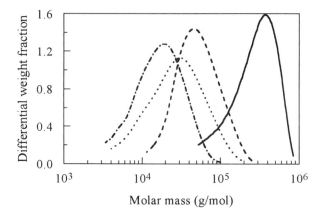

Figure 14. Differential distributions of the CMC 1: no degradation (———), ultrasonic degradation (---), enzymatic degradation (···), combination of ultrasonic and enzymatic degradation (-·-·-).

Polyacrylamide. In addition to the polymers based on renewable raw materials, synthetic polymers also enjoy widespread use. Polyacrylamide is one such polymer. In order to establish structure-property relationships for polyacrylamide, such as the Mark-Houwink equation, homologous series are needed, i.e. polyacrylamide samples which have differing molar masses but identical polydispersities. Such samples are produced with the aid of ultrasonic degradation, the absolute differential distributions of which are shown in Figure 15.

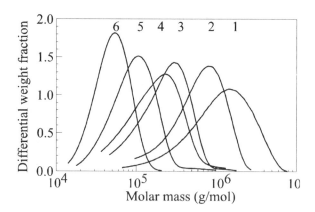

Figure 15. Differential distribution curves obtained by means of SEC/MALLS/ DRI for ultrasonically degraded polyacrylamides.

Peak 1 shows the non-degraded sample, increasing sample number denotes a greater duration of ultrasonic degradation. It can be clearly seen that the molar mass is displaced to smaller values as the length of ultrasonic treatment increases and that the distribution becomes narrower. Table II gives the duration of ultrasonic treatment and the respective weight-average molar masses as well as the polydispersity, M_W/M_n. The values for M_W/M_n fall from 1.8 to 1.2. This is an indication that scission is occurring in the chain centre (see above (24)). However, this change is small in contrast to a change in polydispersity observed in samples where molar masses have been adjusted by different reaction conditions.

Table II. Molar mass and polydispersity values of the ultrasonically degraded polyacrylamide samples. The molar mass distributions were measured by SEC/MALLS/DRI.

Sample	Degradation time (min)	M_W (g/mol)	M_W/M_n (-)
PAAm 1	0	1.5×10^6	1.8
PAAm 2	1	7.5×10^5	1.7
PAAm 3	5	2.8×10^5	1.6
PAAm 4	10	2.0×10^5	1.6
PAAm 5	20	1.1×10^5	1.4
PAAm 6	45	5.7×10^4	1.2

Characterization by means of FFFF/MALLS/DRI

The principle of FFFF was developed at the end of the 60s by Giddings. It is an absolute method because under ideal elution conditions the diffusion coefficient can be determined directly from the retention time. In order to determine the molar mass distribution, either the corresponding D-M relationship is needed or the system has to be calibrated. However, as with SEC, if FFFF is coupled with a light-scattering photometer the molar mass can be determined directly for each fraction. In 1994 Kulicke et al. first succeeded in coupling the FFFF method of fractionation with a light-scattering detector.

As well as polymers dissolved in a molecularly disperse form, FFFF can also be used to characterize particulate systems. Hence with this method it has been possible to fully characterize samples that contain both molecularly disperse and particulate components. It has already been shown that this is a powerful method of characterizing polymers (33) and polyelectrolytes dissolved in a molecularly disperse form, as well as polystyrene latex dispersions (11,34).

Albumin and tobacco mosaic virus. In order to examine the accuracy of a new method of characterization, it is important to have systems that can also be characterized exactly with other methods. For this reason the samples chosen for the investigation were bovine serum albumin (BSA), which as a globular protein may be regarded in idealized form as a sphere, and the rod-shaped tobacco mosaic virus (TMV). Due to its sequence of amino acids the protein BSA has a definite molar mass and size. As does TMV, it occurs in a monodisperse state.

Figure 16 A shows the elution graph of BSA. In addition to the concentration signal (plotted in arbitrary units) and the directly measured molar masses (MALLS/DRI), the figure also shows the diameters calculated via FFFF theory (dotted line). Next to the first intensive peak, a second peak can be seen which accounts for only approx. 13%. The molar mass remains constant in each of the peak ranges. A result of 127,000 g/mol is recorded for the second peak in comparison with 64,500 g/mol, and it may be assumed that this is due to the dimer. The curve of the radii (FFFF theory) has been calculated without taking into account the band broadening and would suggest polydispersity of the BSA. A similar picture is given for the TMV. As can be seen in Figure 16 B, the curve for the radii of gyration also remains constant. The values determined for R_G are 91.7 nm, which agree well with the value of 92.4 nm (*35*) from the literature. These results are already discussed in ref. *36*.

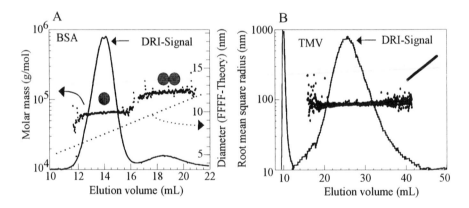

Figure 16.   A: Elution profile, molar masses and hydrodynamic radii from FFFF theory of a bovine serum albumin sample (BSA) measured by FFFF/MALLS/DRI in aqueous sodium dodecyl sulphate solution, containing 0.02% w/w sodium azide, T = 298 K, $\dot{V}_Z$ = 0.5 mL/min, $\dot{V}_X$ = 5 mL/min, M_{mon} = 6.45 × 10⁴ g/mol, M_{dimer} = 1.28 × 10⁵ g/mol.

B: Elution profile and root mean square radii for a tobacco mosaic virus sample (TMV) measured by FFFF/MALLS/DRI in aqueous sodium dodecyl sulphate solution, containing 0.02% w/w sodium azide, T = 298 K, $\dot{V}_Z$ = 2.0 mL/min, $\dot{V}_X$ = 0.45 mL/min, R_G = 91.7 nm.

(Reproduced with permission from reference *36*. Copyright 1998 John Wiley.)

Cellulose derivatives. Cellulose derivatives are used on a large scale industrially. The characterization of these polymers is to some extent complicated by the fact that they contain associations and aggregations.

Figure 17 A shows the concentration profile of an FFFF/MALLS/DRI measurement for an HEC sample. The scattered light signal for the 90° detector has also been included in the graph. The injection is made after 2 mL have been eluted. The stop-flow relaxation begins after the injection delay. The actual sample is only eluted after the void peak, which is caused by that sample content with very low (high D) or very high (steric elution) molar mass. The distribution is, as can be seen in Figure 17 B, relatively narrow ($M_w/M_n = 1.26$).

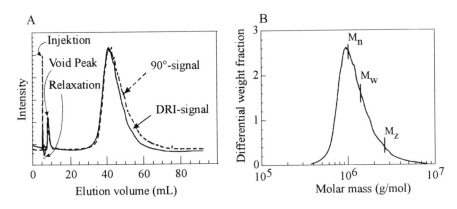

Figure 17. A: Elution profile for hydroxyethyl cellulose sample (HEC) measured by FFFF/MALLS/DRI in aqueous 0.1 M $NaNO_3$ solution, containing 0.02% w/w sodium azide, T = 298 K, $\dot{V}_z$ = 1.0 mL/min, $\dot{V}_x$ = 2.0 mL/min to 0.05 mL/min in 60 min.
B: Differential distribution: $M_n = 1.1 \times 10^6$ g/mol, $M_z = 1.4 \times 10^6$ g/mol, $M_z = 1.9 \times 10^6$ g/mol.

CMC is another cellulose derivative that is put to a wide variety of uses. The elution graph of an FFFF/MALLS/DRI measurement of the CMC 2 can be seen in Figure 18. In order to record the entire sample with the equipment used, the measurements were carried out in PFF mode. It can be seen that sample is eluted over the entire measuring range and that the molar masses ($\sim 3 \times 10^5$ - 2×10^7 g/mol) and radii (20-200 nm) increase continuously. Two regions with different gradients can be distinguished for the CMC 2 in the plot of the radii and diffusion coefficients against the molar masses (Figure 19). The following relationships are given for the molar mass: $M < 10^6$ g/mol: $R_G = 6.2 \times 10^{-3}$ nm $M^{0.70}$
$M > 10^6$ g/mol: $R_G = 0.8$ nm $M^{0.36}$

The ν exponents (reciprocal fractal dimension) reveal that in addition to expanded molecules ($\nu = 0.70$) very compact aggregates ($\nu = 0.36$) are also present.

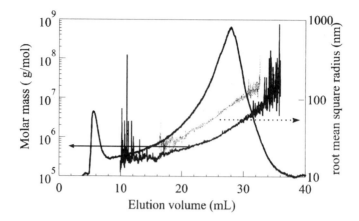

Figure 18. Elution profile, molar masses and r.m.s. radii for carboxymethyl cellulose sample (CMC 2) measured by FFFF/MALLS/DRI in aqueous 0.1 M $NaNO_3$ solution, containing 0.02% w/w sodium azide, T = 298 K, $\dot{V}_z$ = 0.5 mL/min, $\dot{V}_x$ = 0.5 mL/min to 0.05 mL/min in 60 min.

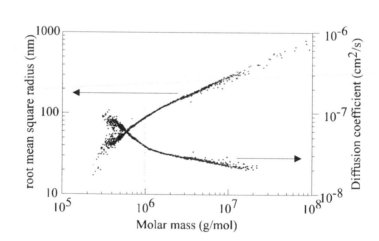

Figure 19. Double logarithmic plot of root mean square radius and diffusion coefficient versus molar mass.

Poly(diallyldimethylammoniumchloride) (poly-DADMAC). Determining the molar mass distributions of cationic polyelectrolytes by means of SEC proves difficult because the charge on the sample may lead to interaction with the stationary phase in the column. FFFF now makes it possible not only to analyse polyanions (see below) but also polycations without altering the structure or performing elaborate changes of solvent. Figure 20 shows the differential molar mass distribution for both a laboratory and an industrial sample of poly-DADMAC.

A relatively narrow distribution is observed for the laboratory sample, whereas the industrial sample is not only considerably more broadly distributed but also displays a shoulder on the high-molar-mass flank (see also Ref. *36*). This indicates a difference in reaction conditions during the synthesis, an observation which was later confirmed by the manufacturer. In the industrial sample fresh initiator was added at a relatively late stage in the reaction.

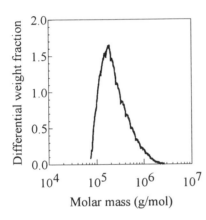 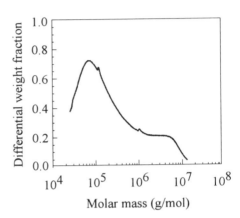

Figure 20. Differential distribution of a laboratory (left) and an industrial (right) sample of cationic polyelectrolyte [poly-(diallyldimethyl-ammonium) chloride] measured by FFFF/MALLS/DRI in aqueous 0.1 M $NaNO_3$ solution, containing 0.02% w/w sodium azide, T = 298 K, lab. Product: $\dot{V}_z$ = 0.5 mL/min, $\dot{V}_x$ = 1.7 mL/min, M_w = 3.4 × 10^5 g/mol, M_w/M_n = 1.7; industrial product: $\dot{V}_z$ = 1.0 mL/min, $\dot{V}_x$ = 2.5 mL/min to 0.1 mL/min in 42 min, M_w = 9.0 × 10^5 g/mol, M_w/M_n = 5.6.

Polystyrene sulphonates. The following aims to demonstrate that with the aid of FFFF it is possible to separate a polystyrene sulphonate mixture composed of seven standards (see Ref. *20*). In addition, the efficiency of the two fractionating units will be compared briefly. Figure 21 contains the elution profile and the corresponding molar masses from an FFFF/MALLS/DRI measurement for a mixture of all seven standards. In this case the analysis was performed with a linearly decreasing cross flow. It can be seen that separation and detection are possible over a very wide range (M_w/M_n = 14).

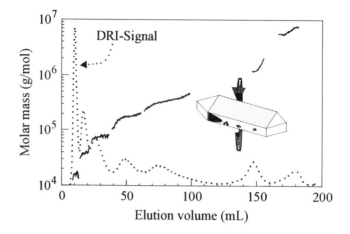

Figure 21. Molar mass obtained by FFFF/MALLS/DRI for a mixture of seven polystyrene samples measured by FFFF/MALLS/DRI in aqueous 0.1 M NaNO$_3$ solution, containing 0.02% w/w sodium azide, T = 298 K; $\dot{V}_z$ = 0.48 mL/min, $\dot{V}_x$ = 6.0 mL/min to 0.17 mL/min in 333 min.

In the attempt to separate this 7-component mixture by means of SEC, it was observed that the separating capacity of the SEC was not sufficient to separate a mixture that was so broadly distributed (Figure 22). Even from the profile of the single measurements (dotted line in Figure 22) it can be seen that separation is not possible. The two high-molar-mass standards are already located at the edge of the upper measuring range and are not ideally eluted, which leads to distortion of the molar masses of the small standards at higher values.

If the two high-molar-mass samples are removed from the mixture, a distinct improvement in the resolution is observed (Figure 23). The individual peaks can be clearly recognized and even the molar masses agree with those of the individual measurements. However, only three of the five standards are baseline separated. This shows that FFFF has a better separating power due to its higher selectivity and a greater peak capacity. The capacity of the SEC can only be raised by installing additional columns, whereas that of FFFF can easily be controlled by the ratio of cross flow to channel flow.

Summary

It has been shown that SEC and FFFF are both powerful methods of separation. In FFFF the separating strength can be easily varied by altering the flow relationships. This method is also superior to SEC in terms of peak capacity and selectivity. However, SEC has the advantage of being easy to use and stable and is thus suitable for standard problems. In combination with detectors for scattered light and concentration, both methods have proven to be ideal tools for determining molar mass

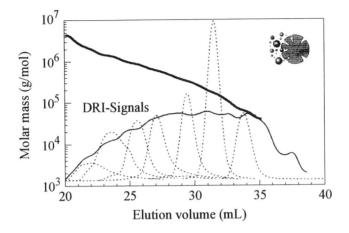

Figure 22. SEC/MALLS/DRI measurement on a mixture of seven NaPSS standards. Apart from the elution graph (continuous line) the measured molar masses and the elution profiles of the individual components (dotted lines) are also plotted.

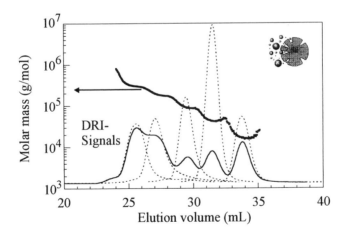

Figure 23. SEC/MALLS/DRI measurement on a mixture of five NaPSS standards. Apart from the elution graph (continuous line), the measured molar masses and the elution profiles of the individual components (dotted lines) are also plotted.

distributions. Consequently, the two techniques should be regarded as complementary rather than competitive.

Glossary

A_2 second virial coefficient in the Zimm-Debye-equation
a   exponent of Mark-Houwink ($[\eta]$-M) equation
c   concentration
D   diffusion coefficient
DRI   differential refractive index
dn/dc   refractive index increment
FFFF   flow field-flow fractionation
K   light scattering constant, equal to $4\pi^2(dn/dc)^2 n_0^2 / N_A \lambda_0^4$
l   equilibrium layer
M   molar mass
$M_{n/w/z}$ number/weight/z- average molar mass
MALLS   multi-angle laser light scattering
n   refractive index
n_0 refractive index of pure solvent
P   degree of polymerisation
$P(\vartheta)$ scattering function
PFF   programmed field of force
R_ϑ excess Rayleigh ratio
SEC   size exclusion chromatography
t_r retention time
V   volume
$\dot{V}_X$ cross flow
$\dot{V}_Z$ channel flow
w   channel height

Greek Characters

δ deflection
ϑ scattering angle
$[\eta]$ intrinsic viscosity
ν exponent of R-M-equation
λ_0 vacuum wavelength of incident light

Acknowledgments

This work was kindly supported by the Deutsche Forschungsgemeinschaft (DFG).
     The authors wish to thank Dr. W. Jaeger, Fraunhofer Institut für Angewandte Polymerforschung [Fraunhofer Institute for Applied Polymer Research], for providing the poly-DADMAC samples and Dr. Koch and Dr. K. Szablikowski [Wolff Walsrode AG] for donating the CMC samples.

References

1. *Römpps Chemie Lexikon*; Frank'sche Verlagshandlung, W. Keller&Co., Stuttgart, Germany, 1987, Vol. 5

2. Kulicke, W.-M. (Ed.) *Fließverhalten von Stoffen und Stoffgemischen*; Hüthig & Wepf, Heidelberg, Germany, 1986, pp 88

3. Müller, M. PCT Patent application, EP 93 / 02375

4. Kulicke, W.-M.; Roessner, D.; Kull, W. *starch/stärke* **1993**, *45*, 217

5. Yau, W. W.; Kirkland, J. J.; Bly, D. D. *Modern Size-Exclusion Liqid Chromato-graphy*; J. Wiley&Sons, New York, 1979

6. Provder, T. (Ed.); ACS 245, 1984

7. Klein, J.; Kulicke, W.-M.; Hollmann, J. *Chromatographie zur Bestimmung der Molmassen-und Teilchengrößenverteilung von Polymeren*; in: Günzler, H. (Ed.) *Analytiker-Taschenbuch* Vol. 19; Springer: Berlin, Germany, 1998, pp 317

8. Giddings J. C. *Sep. Sci.* **1966**, *1*, 123

9. Giddings, J. C.; Yang, F. J.; Myers, M. N. *Anal. Chem.* **1976**, *48*, 1126

10. Giddings, J. C. *Sep. Sci. a. Tech.* **1978**, *13*, 241

11. Roessner, D.; Kulicke, W.-M. *J. Chromator.* **1994**, *A 687*, 249

12. Elias, H. G. *Makomoleküle: Struktur, Eigenschaft, Synthese, Stoffe, Technologie*, Hüthig & Wepf, Heidelberg, Germany, 1981

13. Johnson, P.; Brown, W. in: Harding, S.E.; Sattelle, D. B.; Bloomfield, V. A.: *Laser Light Scattering in Biochemistry*; Royal Soc. of Chem. 1992

14. Reinhard, U. T.; Meyer de Groot, E. L.; Fuller, G. G. *Macromol. Chem. Phys.* **1995**, *196*, 63

15. Pöhlmann R. *Krankenhauspharmazie* **1991**, *11*, 496

16. Förster, H.; Asskali, F.; Nitsch, E. (for Laevosan GmbH) published specification DE 4123000 A1, Germany 1993

17. Elias, H.-G. *Makromoleküle; Vol.2: Technologie*; Hüthig & Wepf, Basel, Germany 1992

18. Lutz H. *Plasmaersatzmittel*, Georg Thieme Verlag, Stuttgart, Germany 1975

19. *Ullmanns Enzyklopädie der Technischen Chemie*; VCH, Weinheim, Germany 1982, Vol. 9, 192

20. Thielking, H.; Kulicke, W.-M. *Anal. Chem.* **1996**, *68*, 1169

21. Kulicke, W.-M.; Klein, J. *Angew. Makromol. Chem.* **1978**, *69*, 169

22. Kulicke, W.-M. in: Houben-Weyl: *Methoden der organischen Chemie - Makromolekulare Stoffe*; Georg Thieme Verlag, Stuttgart, Germany, 1987, Vol. E20; 1176

23. Jaeger, W. ; Gohlke, U.; Hahn, M.; Wandrey, C.; Dietrich, K. *Acta Polymerica* **1989**, *40/3*, 161

24. Kulicke, W.-M.; Otto, M.; Baar, A. *Makromol. Chem.* **1993**, *194*, 751

25. Ryskin, G. *J. Fluid Mech.* **1990**, *218*, 239

26. Odell, J. A.; Muller, J. A.; Narh, K. A.; Keller, A. *Macromolecules* **1990**, *23*, 3092

27. Kulicke, W.-M.; Böse, N. *Colloid & Polymer Sci.* **1984**, *262*, 197

28. Grushka, E.; Caldwell, K. D.; Myers, M. N.; Giddings, J. C. *Separ. Purif. Methods* **1973**, *2*, 127

29. Chen, X.; Wahlund, K.-G.; Giddings, J. C. *Anal. Chem.* **1988**, *60*, 367
30. Förster, H.; Asskali, F.; Nitsch E. (for Laevosan GmbH) published specification DE 4122999 A1, Germany 1993
31. Käuper, P.; Kulicke, W.-M.; Horner, S.; Sake, B.; Puls, J.; Kunze, J.; Fink, H.-P.; Heinze, U.; Heinze, Th.; Klohr, E.-A.; Thielking, H.; Koch, W. *Angew. Makro. Chem.*; in print
32. Kasulke, U.; Linow, K.-J.; Philipp, B.; Dautzenberg, H. *Acta Polymerica* **1988**, *39*, 129
33. Adolphi U.; Kulicke W.-M. *Polymer* **1997**, *38 No. 7*, 1513
34. Thielking, H.; Roessner, D.; Kulicke, W.-M. *Anal. Chem.* **1995**, *67*, 3229
35. Boedtker, H.; Simmons, N. S. *J. Am. Soc.* **1958**, *80*, 2550
36. Thielking, H.; Kulicke, W.-M. *J. Microcolumn Sep.* **1998**, *10*, 51

Chapter 10

Cross-Fractionation of Copolymers Using SEC and Thermal FFF for Determination of Molecular Weight and Composition

Sun Joo Jeon and Martin E. Schimpf[1]

Department of Chemistry, Boise State University, 1910 University Drive, Boise, ID 83725

The detailed characterization of polymer and copolymer mixtures requires the determination of both molecular weight and chemical composition. Such information can be obtained from the combination of size-exclusion chromatography (SEC) and thermal field-flow fractionation (ThFFF) with mass and viscosity detectors. In the first step of the method, polymer standards and the concept of universal calibration are used to define the dependence of retention on the diffusion coefficient D in an SEC column. Next, the SEC column is used to separate a polymer mixture into elution slices that each have a unique and definable D value. These slices are collected and individually cross-fractionated according to chemical composition by ThFFF. Values of D are combined with measurements of intrinsic viscosity to yield the molecular weight of the resulting fractions. The D value of a separated fraction is combined with its ThFFF retention parameter to yield an associated thermal diffusion coefficient, from which the fraction's chemical composition is obtained. The method is demonstrated with blends of polystyrene-ethylene oxide copolymers and their corresponding homopolymers.

The separation and subsequent analysis of polymer blends and copolymers is a challenge for polymer scientists because of the overlapping effects of molecular weight and chemical composition. Size-exclusion chromatography (SEC), which is typically used to measure molecular weight (M), actually separates polymers according to differences in the diffusion of components into the pores of the SEC packing material (1). The diffusion is governed solely by the hydrodynamic volume

---

[1]Corresponding author.

(V_h) of the components. The ability of SEC to measure M is rooted in the correlation of V_h to the product M[η], where [η] is the intrinsic viscosity of the dissolved polymer. Since the dependence of [η] on M varies with polymer composition, an SEC column that is calibrated with one type of polymer cannot be used to determine values of M for another type of polymer unless [η] is also measured (2). Thus, by calibrating retention volume in an SEC column to the product [η]M rather than simply M, values of M can be determined for a wide range of homopolymers using a single calibration curve. For analyzing polymer blends and copolymers, however, the utility of SEC is diminished by the fact that components differing in both molecular weight and chemical composition coelute when they have the same hydrodynamic volume.

Thermal field-flow fractionation (ThFFF) is another separation tool used to characterize the molecular weight of polymers (3). Its range of applicability complements that of SEC. Thus, ThFFF has virtually no upper molecular weight limit, but its resolving power diminishes below some threshold value of M (4). The threshold value of M varies with chemical composition but it is generally around 10^4 g-mol^{-1}. By contrast, SEC can resolve low molecular weight oligomers but problems arise in the analysis of polymers with M values above 10^6 g-mol^{-1} (5). Like SEC, retention in ThFFF depends on V_h and not on M directly, therefore two polymer components that differ in both chemical composition and molecular weight may coelute. However, ThFFF retention is more strongly influenced by chemical composition. The direct dependence of retention on composition is a unique feature of ThFFF that is not present in SEC (6). As a result, the component combinations that coelute in ThFFF are different than those that coelute in SEC, and the two techniques can be combined to achieve better resolution of complex polymer mixtures than either technique alone.

In this work, we first use SEC to separate the components of a mixture according to differences in V_h. Slices of the resulting elution profile are subsequently cross-fractionated by ThFFF. Since components in the SEC elution slices are nearly homogeneous in V_h, the ThFFF separation is based primarily on differences in chemical composition. As a result, the two-dimensional separation is orthogonal in V_h and chemical composition. Molecular weight information on the separated components is extracted using mass and viscosity detectors, while compositional information is extracted from the dependence of ThFFF retention on polymer composition, which is described next.

A detailed description of the retention mechanism in ThFFF can be found in the literature (7). The dependence of retention on both V_h and chemical composition arises from the nature of the separation mechanism, which involves the balance of two transport processes. The primary transport process consists of the movement of mass in response to a temperature gradient. This process is referred to as thermal diffusion, and varies greatly with the chemical composition of both polymer and solvent. Opposing the motion of thermal diffusion is ordinary (mass) diffusion. Thus, the dependence of ThFFF retention on M, like SEC, stems from a more fundamental dependence of retention on diffusion.

Ordinary diffusion is quantified by the mass diffusion coefficient (D), while thermal diffusion is quantified by the thermal diffusion coefficient (D_T). Retention in

ThFFF can be related to the ratio D_T/D, which is referred to as the Soret coefficient. Because the relationship is well defined, the Soret coefficient of a separated component can be calculated directly from its ThFFF retention parameter; calibration with polymer standards is not required. If an independent measure of D is available (e.g. using SEC or dynamic light scattering), the D_T value of the separated component can be calculated. Furthermore, if the dependence of D_T on chemical composition is known, then compositional information can be obtained. Although a general model for relating D_T to physicochemical parameters of the polymer is not yet available, it is possible to establish the dependence of D_T on copolymer composition empirically. Establishing such relationships is simplified by the fact that D_T is independent of molecular weight (8,9). As a result, only one relationship between D_T and composition needs to be established for an entire class of copolymers, for example all copolymers of styrene and ethylene oxide.

When the dependence of D_T on copolymer composition was first demonstrated (10), two observations were made: (1) for random copolymers, D_T can be described by the weighted average of the D_T values of the corresponding homopolymers, where the weighting factors are the mole fractions of each component in the copolymer; (2) in block copolymers, where the monomeric units (mers) are capable of radial segregation within the dissolved polymer-solvent sphere, the value of D_T is governed by mers located in the outer free-draining region. Radial segregation of mers can occur due to bonding constraints in highly branched block copolymers (11) or as a result of solvent effects (10,12). Solvent effects occur when a block copolymer is dissolved in a solvent that is much better for one of the component blocks. Studies have demonstrated that when a non-selective solvent is used, that is a solvent which is equally good for all copolymer blocks, the dependence of D_T on composition follows the same predictable pattern as that in random copolymers.

Using copolymers of styrene and isoprene, we previously demonstrated the ability to obtain both molecular weight and compositional information by combining ThFFF with measurements of intrinsic viscosity (13). The method utilizes the dependence of D on [η] and the viscosity average molecular weight (M_V) to cast the ThFFF retention parameter in terms of D_T, [η], and M_V. After measuring both the retention parameter and [η] of a copolymer sample in multiple solvents, a series of equations are solved simultaneously to obtain values of M_V and D_T (and from D_T, the composition) for the sample.

The value of cross-fractionating polymer blends using ThFFF and SEC was first demonstrated by van Asten et al. (14). They measured values of D_T to qualitatively identify differences in composition across the SEC elution profile. By using a non-selective solvent in the work described here, we are able to convert measured values of D_T into quantitative information on the chemical composition of polymer and block copolymer blends. In addition, we use a viscosity detector to obtain molecular weight information on the resolved components. Our procedure can be summarized as follows: (1) a mixture of solvents is found that is non-selective for copolymers of styrene and ethylene oxide, and the linear dependence of D_T on copolymer composition is established; (2) the SEC column is calibrated in terms of D versus retention volume using PS standards whose D-values have been characterized by independent measurements; (3) blends of poly(styrene-ethylene oxide) (PSEO) block copolymers and their corresponding homopolymers are separated according to hydrodynamic volume by SEC; (4) SEC fractions, which contain multiple components varying in both molecular weight and composition, are cross-fractionated according to chemical composition by ThFFF; (5) [η] values are measured on the eluting material with a viscosity detector; (6) the values of [η] are combined with D values obtained from the SEC retention volumes to calculate values of M; (7) the D values are combined with ThFFF retention parameters to calculate the D_T values of the eluting ThFFF fractions; (8) the values of D_T are converted into values of chemical composition using the relationship established in step (1).

Theory

In ThFFF, the fundamental retention parameter (λ) is related to the temperature drop across the channel (ΔT) and the transport coefficients by

$$\lambda = \frac{D}{D_T \Delta T} \tag{1}$$

Parameter λ is also related to the volume of liquid (V_r) required to elute a polymer component

$$R = \frac{V^o}{V_r} = 6\lambda\left[\coth(2\lambda)^{-1} - 2\lambda\right] \tag{2}$$

where R is termed the retention ratio and V^o is the geometric volume of the channel. Equations 1 and 2 are actually approximations that neglect perturbations due to the dependence of the carrier liquid viscosity and transport coefficients on temperature. Although corrections to eqs 1 and 2 have been developed to account for such perturbations (15-17), they cannot be used in this work because the required parameters are unavailable. Van Asten et al. (17) showed that in theory, errors in D_T (and therefore composition) associated with neglecting temperature corrections can be as large as 4%. In practice, however, this error is canceled by the use of calibration curves that relate polymer composition to D_T-values calculated without the temperature correction.

In SEC, retention is governed by the polymer's mass diffusion coefficient (D), which is related to the product [η]M as follows (18):

$$D = \frac{kT}{6\pi\eta_o}\left(\frac{10\pi N_A}{3[\eta]M}\right)^{1/3} \tag{3}$$

Here k is Boltzmann's constant, T is temperature, N_A is Avogadro's number, and η_o is the viscosity of the solvent. For a polydisperse polymer, D in eq (3) is an average value for the different molecular weight components and M is the viscosity-average molecular weight (M_V). In the "universal calibration" of SEC columns, a plot of log [η]M versus retention volume (V_r) is established with polymer standards of known molecular weight. Such plots are called universal because they can be used to characterize values of M (or M_V) for polymers of different composition, provided an independent measure of [η] is available (2). However, it is apparent from eq 3 that SEC columns can also be "universally" calibrated in terms of log D versus V_r. Once such a plot is established, D values can be obtained for a wide range of polymers from their values of V_r in the calibrated column. The validity of this approach was demonstrated previously (9).

In the cross-fractionation of polymers by SEC and ThFFF, the D values associated with each SEC fraction are used in two ways: to calculate M from eq 3, and to calculate D_T from eq 1. For precise calculations, the value of D obtained from the SEC calibration curve is first corrected for the difference in temperature between that used in the SEC experiment and that experienced by the polymer component in the ThFFF experiment. This correction is given by

$$D_{ThFFF} = D_{SEC} \frac{\eta_{o,SEC} T_{SEC}}{\eta_{o,cg} T_{cg}} \qquad (4)$$

where the subscript "SEC" refers to values associated with the temperature of the SEC experiment, and the subscript "cg" refers to values associated with the temperature in the ThFFF channel where the center of gravity of the eluting polymer zone is located. Parameter T_{cg} is related to the channel's cold-wall temperature (T_c) by

$$T_{cg} = T_c + \lambda \Delta T \qquad (5)$$

For a copolymer containing two components A and B, the mole-% of component A.(X_A) is calculated from the copolymer's thermal diffusion coefficient ($D_T^{copolymer}$) as follows:

$$X_A = \frac{D_T^{copolymer} - D_T^{B}}{D_T^{A} - D_T^{B}} \times 100 \qquad (6)$$

Here D_T^A and D_T^B are the thermal diffusion coefficients of homopolymers composed of pure A and pure B, respectively. Eq 6 is a consequence of the linear dependence of $D_T^{copolymer}$ on X_A with boundaries defined by D_T^A and D_T^B.

In choosing the width of fractions to be collected from the SEC separation for cross-fractionation by ThFFF, we used the method outlined by van Asten et al. (14). In this method, fraction widths are chosen on the basis of producing the maximum amount of information from the second (ThFFF) dimension, which requires the spread of D values within a given fraction (ΔD) to have a negligible effect on the ThFFF elution profile. The fraction width is chosen such that ΔD will increase the standard deviation of the ThFFF elution profile by a maximum of 10% beyond that produced by nonequilibrium dispersion, which dominates band broadening in ThFFF (20). The maximum value of ΔD can be approximated by (14):

$$\Delta D = 1.3 wR\left(\lambda D / t^\circ\right)^{1/2} \tag{7}$$

where w is the channel thickness and t° is the void time, that is the time required to elute a component that is not affected by the temperature gradient. From the dependence of the SEC retention volume on D, the fraction width in volume units can be calculated. For this work, a fraction width of 250 µL satisfies the requirement for all components analyzed.

Experimental

The SEC column used in this work is an ultrastyragel HT column (Waters Corp., Milford, MA) with a length of 30 cm and an internal diameter of 7.8 mm. The particle diameter is 10 µm and the pore size is 10^4 Å. The column temperature was maintained at 35 °C with a column oven (Timberline Instruments, Boulder, CO). The flow rate of the mobile phase through the SEC column was 0.5 mL/min.

The ThFFF system has been described previously (19). The channel has a thickness of 102 µm, a breadth of 1.9 cm, and a tip-to-tip length of 46 cm. The corresponding void volume is 0.88 mL. A ΔT value of 45 ± 2 °C was maintained with a cold wall temperature of 38.5 ± 0.5 °C. A stop-flow period of 1 minute was used to relax the sample to the cold wall after injection (20). Unless otherwise stated, the flow rate of the carrier liquid in the ThFFF channel was 0.1 mL/min.

The polymers used in this study are summarized in Table I. They consist of polystyrene (PS) obtained from Pressure Chemical Co. (Pittsburgh, PA), polyethylene oxide (PEO) from Scientific Polymer Products, Inc. (Ontario, NY), and several polystyrene-co-ethylene oxide block copolymers (PSEO) from Polymer Laboratories (Amherst, MA). The carrier liquid for the cross-fractionation experiments was a 5:1 (vol/vol) mixture of tetrahydrofuran (THF) and N,N-dimethylformamide (DMF). Carrier liquids were delivered with a Model P-500

Table I. Summary of Polymers used in Cross-Fractionation

Polymer Code	M^a (g-mol^{-1})	Styrene Content[b] (mol-%)	Ethylene Oxide Content[b] (mol-%)	Nominal Polydispersity
PS400	400,000	100	0	<1.06
PEO101	101,000	0	100	1.04
PEO190	190,000	0	100	1.04
PSEO51-85	51,300	84.9	15.1	1.07
PSEO68-5	67,800	5.1	94.9	1.11
PSEO250-70	250,000	69.7	30.3	1.15

[a] determined by supplier using SEC
[b] determined by supplier using NMR

(Pharmacia Biotech, Sweden) syringe pump when viscometric detection was used. The use of a syringe pump minimizes pressure pulses, which cause oscillations in the baseline. Otherwise, a Model 590 high pressure pump (Waters Corp., Amherst MA) was used for solvent delivery. A Model 100 Differential Viscometer (Viscotek Corp., Houston, TX) was used to measure the intrinsic viscosity of eluting polymer fractions. The viscometer data was processed using Unical version 4.07 software from Viscotek. The mass of the injected polymer sample, which is required for the calculation of intrinsic viscosity, was measured by eluting the polymers through a Model 950/14 evaporative light scattering (ELS) detector (Polymer Laboratories, Amherst, MA) with a furnace temperature of 60 °C and an air pressure of 30 psi. The response of the ELS detector was calibrated to polymer mass with PS, PEO, and PSEO standards. Variations in the response factor of the ELS detector with molecular weight and composition are discussed in the Results section. For cross-fractionation, 250 μL fractions were taken directly from the SEC column effluent and collected in 1 mL glass vials. From these fractions, 20 μL aliquots were injected into the ThFFF channel. For reinjection of fractions into the SEC column (discussed further below), 65 μl aliquots were used.

Results and Discussion

Choice of polymer mixtures, solvent, and sample load. Three different mixtures of PS, PEO, and PSEO were prepared. These mixtures are summarized in Table II. For mixture 1, two components were chosen that are resolved by SEC but not ThFFF. Mixture 2 contains two component that coelute with SEC but are resolved by ThFFF. Mixture 3 is the most difficult sample, containing four components that cannot be completely resolved by either SEC or ThFFF alone. All mixtures were prepared from equivalent masses of the included components.

Table II. Summary of Polymer Blends

Mixture Number	Components
1	PEO101 PSEO51-85
2	PEO190 PSEO250-70
3	PS400 PEO101 PSEO51-85 PSEO250-70

Our choice of carrier liquid was based on several requirements: (1) a solvent strength that is sufficient to prevent the interaction of polymer with the SEC packing material; (2) resolution of the ThFFF elution profiles from the void peak (ThFFF retention varies significantly with solvent composition); and (3) D_T values that vary linearly with the composition of the copolymers. Although benzene resulted in adequate ThFFF retention, the PEO samples interacted with the SEC packing material. Therefore, we switched to the more polar solvent tetrahydrofuran (THF). In pure THF, D_T values are not linearly related to composition, as illustrated in Figure 1a. In fact, D_T values are independent of composition in copolymers containing 70-100 mol-% styrene. This behavior is typical of a selective solvent, in this case one that is a better solvent for the styrene blocks compared to the ethylene oxide blocks. The difference in solvating power results in the segregation of styrene segments to the free-draining region of the dissolved polymer, which is the region that dominates thermal diffusion behavior. In order to increase the solvating power for the ethylene oxide segments, DMF was added to the carrier liquid. With a ratio of 5:1 (v/v) THF/DMF, D_T values are linearly related to copolymer composition, as illustrated in Figure 1b. Although we found that higher ratios of DMF (up to 50 vol-%) also yielded a linear dependence of D_T on copolymer composition, ThFFF retention levels decreased. The ratio of 5:1 maximized the level of retention while maintaining a linear dependence of D_T on composition.

Another issue that must be addressed in cross-fractionation experiments is the amount of polymer injected in the first dimension. Sample dilution occurs in both SEC and ThFFF, and the amount injected in the first dimension must be high enough for adequate detection in the second dimension. If the concentration is too low, compositional information obtained from the ThFFF separation is compromised. Unfortunately, the sample load cannot be raised without limit. When the load is too high, the SEC elution profiles become distorted and resolution of the components is diminished. In this work, the initial sample load was varied to identify the maximum amount of sample that could be injected without distortion of the elution profiles. Based on those studies, we chose to inject 65 μL of sample with a total polymer concentration of 1.5 mg/mL.

SEC calibration. The SEC column was calibrated using a series of nearly monodisperse PS standards ranging in molecular weight from 2500 to 4,100,000 g-mol^{-1}. The plot of log D versus retention volume (V_r) for these standards in THF is illustrated in Figure 2. The D values used to establish this plot were calculated from the nominal molecular weight using the following relationship, which was established in THF using a capillary viscometer (21):

$$D = 4.51 \times 10^{-4} \text{ cm}^2\text{-s}^{-1}\text{M}^{-0.575} \tag{8}$$

The plot of log D versus V_r is linear for molecular weights between 10,000 and 900,000 g-mol^{-1}, or D values between 2×10^{-6} and 2×10^{-7} cm^2-s^{-1}. A least-squares fit yields the following relationship:

$$\log D = 0.2374 \, V_r - 8.0472 \tag{9}$$

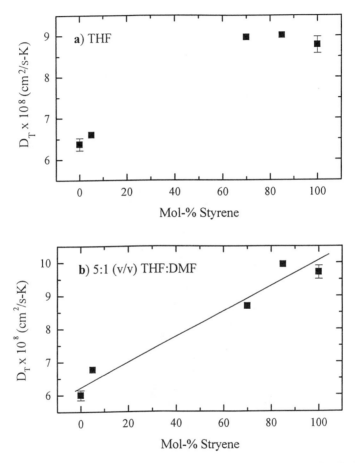

Figure 1.   Dependence of the thermal diffusion coefficient on composition for PSEO copolymers in a) tetrahydrofuran and b) a 5 : 1 (v/v) mixture of tetrahydrofuran and N,N-dimethylformamide. The error bars represent 1 standard error based on the measurement of several polymers with different molecular weights.

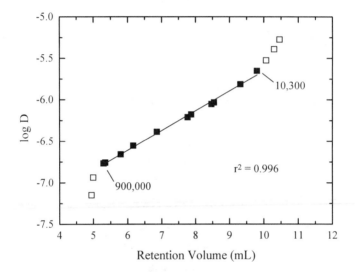

Figure 2.   Plot of log D versus SEC retention volume for PS homopolymers. Values of D were calculated from the nominal molecular weight using eq 8.

We note that eq 9 was established in pure THF, while the carrier liquid used for cross-fractionation was a mixture of 5:1 THF/DMF. This is because the dependence of D on M for PS in the mixed solvent is not readily available. As a result, D values calculated from their SEC retention in 5:1 THF/DMF using eq 9 will contain an undetermined amount of systematic error if swelling of the SEC packing material is different in pure THF compared to 5:1 THF/DMF. Such errors, which propagate to calculations of molecular weight, are considered below in discussing calculated values of M_V in the cross-fractionation experiments. For now, we note that the results indicate minimal effect. Such uncertainties can be avoided altogether if an absolute measure of D is available in the solvent mixture required for cross-fractionation. Measurement of D could be made by dynamic light scattering, for example, but this is not available in our laboratories. Finally, we note that in contrast to molecular weight, copolymer composition is not affected by systematic errors in D. As we demonstrated in a previous work (13), systematic errors in D (or $[\eta]$) are not transferred to calculations of X_A because they are canceled by an identical error in D_T during the calibration of D_T to X_A.

Cross-fractionation. Figure 3a illustrates the separation by SEC of mixture 1, which contains equal amounts of the homopolymer PEO101 (M = 101,000) and the copolymer PSEO51-85 (M = 51,000; 84.9 mol-% styrene). The two components are nearly baseline resolved. Because the individual components are nearly monodisperse, each elutes within a relatively narrow range of V_r. In fact the width of the resolved SEC peaks in Figure 3a is dominated by band broadening rather than sample polydispersity. The dominance of band broadening is confirmed by reinjection of the collected fractions, which are marked by vertical lines in Figure 3a. The SEC elution profiles of these reinjected fractions are displayed in Figure 3b. Note that although the mean values of V_r for fractions f1-2 and f1-3 differ by 250 μL, the difference is reduced to 100 μL upon reinjection. Thus, the initial difference of 250 μL is primarily due to band broadening rather than the separation of components differing in D.

The shift in V_r with reinjection has significant consequences for the cross-fractionation procedure. The accuracy of D_T values calculated from ThFFF retention parameters can be no better than the accuracy of the measured D values. However, Figure 3 demonstrates that when band broadening is significant, one cannot simply assign a D value to an elution slice based on its mean V_r value (or even the value associated with the slice's center of gravity). Instead, one must reinject the collected fraction to obtain a more accurate value of D. Even when the profile is not dominated by band broadening, one must be careful in assigning a D value to elution slices in the tail regions of the profile, where band broadening is more significant. In this work, all D values associated with SEC fractions were calculated from their elution-peak maxima upon reinjection. We note, however, that for polydisperse samples, van Asten et al. (14) demonstrated that the peak maxima of reinjected fractions match the mean retention volumes associated with the original elution profile, so that reinjection is not required. Of course, other methods (such as dynamic light scattering) could be used to obtain an independent measurements of D on each fraction.

a) initial fractionation

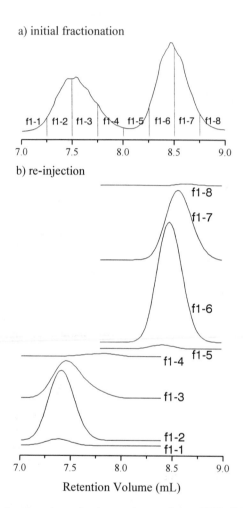

b) re-injection

Retention Volume (mL)

Figure 3. SEC fractionation of mixture 1 containing PEO (M = 101,000) and PSEO51-85 (M = 51,300; 84.9 mol-% styrene). a) Initial fractionation; the slices collected for reinjection and cross-fractionation are delinated by vertical lines. b) SEC analysis by reinjection of the sample slices from a).

Figure 4 illustrates the ThFFF elution profiles obtained on fractions collected from the SEC separation of mixture 1. In principle, these profiles could be directly converted into a distribution of chemical composition because D is approximately constant across the profile. However, since the fractions contain relatively monodisperse components, band broadening rather than differences in composition dominates the profiles. Therefore, it is more meaningful to simply calculate an average composition for each SEC fraction rather than a distribution. These values are summarized in Table III; we note that several fractions are missing from the table because they were too dilute for adequate detection in the ThFFF experiment. As expected for this simple case, where resolution of the two components is nearly complete, we have good agreement between the calculated and nominal values. The agreement indicates that the D values calculated from SEC are valid, even though the solvent used for calibration of the SEC column (pure THF) is slightly different than that used in the cross-fractionation procedure.

For mixture 1, resolution of the two components was achieved in the first separation dimension, that is by SEC. In this case, cross-fractionation by ThFFF does not serve to further resolve the components but merely to characterize the

Table III. Measured Values of Chemical Composition and Comparison with Nominal Values

Fraction	$D_T \times 10^8$ $(cm^2/s\text{-}K)$	Mol-% Styrene experimental	nominal	Deviation (%)
f1-2	6.61	3.0	0.0	3.0
f1-3	6.78	7.6	0.0	7.6
f1-6	9.54	87.5	84.9	2.6
f1-7	9.5	86.3	84.9	1.4
f2-1	6.40	-3.3	0.0	-3.3
	9.01	72.2	69.7	2.5
f2-2	6.47	-1.1	0.0	-1.1
	8.94	70.0	69.7	0.3
f2-3	6.58	2.0	0.0	2.0
	8.63	60.7	69.7	-9.0
f3-2	10.02	100.5	100.0	0.5
f3-3	10.02	101.4	100.0	1.4
f3-4	9.01	78.6	69.7	8.9
f3-5	9.26	71.9	69.7	2.2
f3-7	6.51	-0.1	0.0	-0.1
f3-8	6.64	3.8	0.0	3.8
f3-11	9.57	87.6	84.9	2.7
f3-12	9.54	86.9	84.9	2.0

a) SEC fractionation

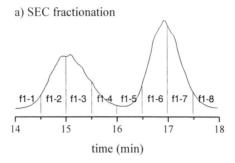

b) ThFFF cross-fractionation

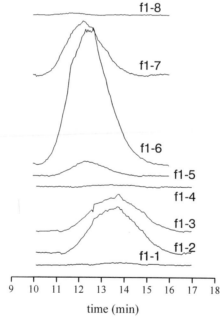

Figure 4. Cross-fractionation of mixture 1 containing PEO (M = 101,000) and PSEO51-85 (M = 51,300; 84.9 mol-% styrene). a) SEC fractionation. b) ThFFF analysis of the sample slices delineated by vertical lines in a).

composition of each component. As a result, the molecular weight of the two components in mixture 1 can be measured by attaching the viscometer to either dimension. However, several factors lead to more accurate results when the molecular weight is measured in the first dimension. First, the signal-to-noise ratio of the viscometer is much greater in the first dimension. Sensitivity is a significant issue with viscometric detectors, especially for molecular weights below 10^5 g-mol^{-1}. Perhaps equally important is the fact that the mass injected in the first dimension is known with high precision since the initial sample solution is prepared by gravimetric procedures. An accurate sample mass is critical to the calculation of intrinsic viscosity. Thus, the Unical software requires the user to input both the sample load and the injection volume, from which it calculates the concentration associated with each digitized point in the elution profile using the signal from the mass detector. In this calculation, the software assumes a linear dependence of detector response on concentration, but the explicit dependence is not required. (The calculation also assumes full recovery of the sample.) When the viscometer is placed in the second dimension, the user must calculate the injected sample mass from the detector signal of the appropriate SEC elution slice. This calculation requires specific knowledge of the dependence of response on concentration. That dependence may change with copolymer composition, and in case of an ELS detector, with molecular weight as well. Although variations in the detector response with composition and molecular weight can be established, as we demonstrate below, the extra step results in a less precise value of the sample load and therefore an additional source of uncertainty in the molecular weight calculation.

Values of M_V calculated on the SEC fractions are summarized in Table IV. Only the two center fractions of each component in mixture 1 were analyzed; fractions in the tail regions could not be characterized due to the lack of an adequate signal from the viscosity detector. The agreement between calculated and nominal molecular weights is generally quite good. The larger error (14.8%) contained in fraction f1-6 is probably due to incomplete resolution. No attempt was made to improve resolution by optimizing the parameters of the SEC experiment.

The cross-fractionation of mixture 2 is illustrated in Figure 5. Note that the SEC separation results in a single symmetrical peak. Only after separation by ThFFF does the bimodal nature of the mixture become evident. We note that compared to mixture 1, the mean values of V_r for the fractions in mixture 2 were more closely matched upon reinjection. This is expected because the contribution of sample polydispersity to the width of the elution profile is more significant. Nevertheless, reinjection was still used to increase the accuracy of the calculated D values, and consequently the accuracy of the compositional information.

Although two components are visible in the ThFFF fractogram of mixture 2, their resolution is incomplete. The resolution could undoubtedly be improved with a more contemporary channel, which is capable of sustaining both a lower cold-wall temperature and a higher field strength (22). Fortunately, the resolution here is sufficient to identify two peaks, from which the composition of the two components can be calculated; these are summarized in Table III. The calculated compositions match the nominal values well. Fraction f2-3 contains the largest error (9%) because it was the least resolved by ThFFF.

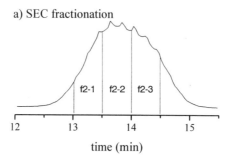

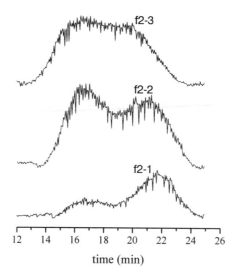

Figure 5. SEC fractionation of mixture 2 containing PEO190 (M = 190,000) and PSEO250-70 (M = 250,000; 69.7 mol-% styrene). a) SEC fractionation. b) ThFFF analysis of the sample slices delineated by vertical lines in a).

Table IV. Measured Values of Molecular Weight and Comparison with Nominal Values

Fraction	[η] (dL/g)	Molecular Weight experimental	[a]nominal	Deviation (%)
f1-2	0.684	101000	101000	0
f1-3	0.657	100000	101000	-1.0
f1-6	0.243	58900	51300	14.8
f1-7	0.261	52200	51300	1.8
f2-1	1.180	173000	190000	-8.9
	0.855	239000	250000	-4.4
f2-2	0.941	181000	190000	-4.7
	0.770	222000	250000	-11.7
f2-3	1.129	142000	190000	-25.3
	0.906	177000	250000	-29.2
f3-2	1.283	418000	400000	4.5
f3-3	1.368	379000	400000	-5.3
f3-4	1.009	248000	250000	-0.8
f3-5	0.906	220000	250000	-12.0
f3-7	0.720	96300	101000	-4.7
f3-8	0.630	106000	101000	5.0
f3-11	0.270	50500	51300	-1.6
f3-12	0.223	55300	51300	7.8

[a] viscosity-average

Since the components of mixture 2 were not resolved by SEC, the molecular weights were determined in the second dimension, that is by attaching the mass and viscometry detectors to the ThFFF channel. In order to increase the sensitivity of the viscometer, the flow rate was increased to 0.2 mL/min for those determinations. As we stated above, calculation of the injected mass requires calibration of the ELS detector response to sample load. The dependence of peak area on sample mass was established from ThFFF elution profiles obtained on several samples across the range of copolymer compositions, each prepared with a concentration of 1 mg/mL. The detector sensitivity was found to change with styrene content. A plot of the relative response of the detector versus styrene composition is illustrated in Figure 6. The line in this plot is a least-squares fit of the data, and was used to obtain the appropriate calibration constant for determining the sample load, once the composition of the component was established.

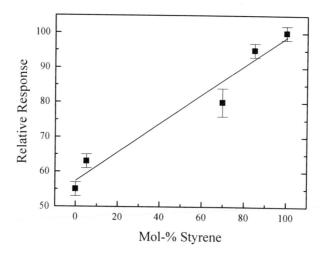

Figure 6. Plot of the relative response of the ELS detector versus copolymer composition. The line is a least-squares fit and was used to adjust the mass injected into the ThFFF channel as calculated from the area under the SEC elution profile.

We also considered variations in the response of the ELS detector with molecular weight. Fortunately, we found the response to be independent of molecular weight for homopolymers of both PS and PEO over the molecular weight range examined in this work. However, the reader should be aware that if the method is applied to other copolymers, or even to PSEO copolymers having a broader molecular weight range than that used in this work, the dependence of detector response on molecular weight must be checked, especially when an ELS detector is used.

The calculated values of M_V for mixture 2 are summarized in Table IV. The calculated values are lower than the nominal values in each case, indicating a systematic error in the method. Such error is most likely propagated from errors in the sample load, which are calculated from the ELS detector signal. Although beyond the scope of this work, the systematic error could be reduced by injecting more samples with different concentrations for an improved calibration of the detector response, rather than relying on a single concentration of 1 mg/mL for each of the different compositions.

The cross-fractionation of mixture 3 is illustrated in Figure 7. Although four components are indicated, the first two components to elute in the SEC separation are only partially resolved. By cross-fractionating the sample with ThFFF, resolution of the first two components is clearly enhanced. On the other hand, the second two components, which are well separated by SEC, elute at nearly the same time in ThFFF. Thus, better resolution of all four components is achieved by combining SEC and ThFFF compared to either technique alone. It is interesting that the ThFFF elution profile of fraction f3-4, which is contaminated with a small amount of PS400, does not contain two resolved peaks, as expected. Instead, a single peak appears with a mean V_r value near that of pure PSEO250-70, but shifted slightly toward that of pure PS400. The reason for this is not clear.

The molecular weights calculated on the components of mixture 3 (Table IV) are generally within about 5% of the nominal values. The most notable exception is fraction f3-5, with an error of 12%. The calculated compositions of fractions f3-1 through f3-3 (Table III) are close to 100% PS, indicating a PS homopolymer. The calculated composition of fraction f3-5 (71.9 mol-% styrene) is close to the nominal value of PSEO250-70 (69.7 mol-% styrene). However, the calculated composition for fraction f3-4 (78.6 mol-% styrene) is closer to the nominal value for PSEO51-85 (84.9 mol-% styrene). From individual fractograms of the pure components, we know that the primary component in fraction f3-4 is in fact PSEO250-70. The large error in calculated composition for this fraction is probably due to incomplete resolution. Compositions calculated on the remaining fractions agree well with the nominal values.

In general, errors in composition are below 3-4%, and we believe from the limited data presented here that differences of 10% in styrene content are easily detected by the method. Furthermore, based on measurements of the homopolymer fractions, we believe the method capable of detecting contents of either styrene or ethylene oxide as low as 10%.

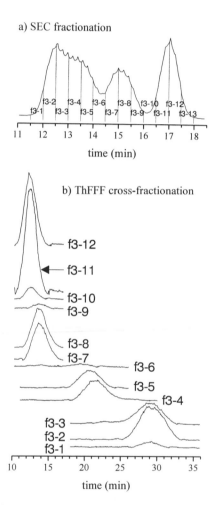

a) SEC fractionation

b) ThFFF cross-fractionation

Figure 7.   SEC fractionation of mixture 3 containing PS400 (M = 400,000), PEO101 (M = 101,000), PSEO51-85 (M = 51,300; 84.9 mol-% styrene), and PSEO250-70 (M = 250,000; 69.7 mol-% styrene). a) SEC fractionation. b) ThFFF analysis of the sample slices delineated by vertical lines in a).

Literature Cited

1. Mourey T. H.; Schunk, T. C. In *Chromatography*; Heftmann, E., Ed., J. Chromatogr. Library Vol. 51B; Elsevier, Amsterdam, 1992; Chapter 22.
2. Williams, D. L.; Pretus, H. A.; McNamee, R.; Jones, E. In *International GPC Symposium '91 Proceedings;* Waters, Division of Millipore Corp.: Newton, MA, 1991.
3. Schimpf, M. E. *Tr. Polym. Sci.* **1996**, *4*, 114-121.
4. Sisson, R.; Giddings, J. C. *Anal. Chem.* **1994**, *66*, 4043-4053.
5. Lee, S.; Kwon, O.-S. In *Chromatographic Characterization of Polymers: Hyphenated and Multidimensional Techniques*, Provder, T.; Barth, H. G.; Urban, M. W., Ed., Advances in Chemistry Series 247; American Chemical Society, Washington, DC, 1995; pp. 93-108.
6. Gunderson, J. J.; Giddings, J. C. *Macromolecules* **1986**, *19*, 2618.
7. Schimpf, M. E. *J. Chromatogr.* **1990**, *517*, 405.
8. Schimpf, M. E.; Giddings, J. C. *Macromolecules* **1987**, *20*, 1561.
9. Schimpf, M. E.; Giddings, J. C. *J. Polym. Sci., Polym. Phys. Ed.* **1989**, *27*, 1317.
10. Schimpf, M. E.; Giddings, J. C. *J. Polym. Sci., Polym. Phys. Ed.* **1990**, *28*, 2673.
11. Schimpf, M. E., Rue, C. A.; Mercer, G.; Wheeler, L. M.; Romeo, P. F. *J. Coatings Tech.* **1993**, *65*, 51-56.
12. Schimpf, M. E.; Wheeler, L. M.; Romeo, P. F. In *Chromatography of Polymers*; Provder, T., Ed.; ACS Symposium Series No. 521; American Chemical Society; Washington DC, 1993, Chapter 5.
13. Schimpf, M. E., In *Chromatographic Characterization of Polymers: Hyphenated and Multidimensional Techniques*, Provder, T.; Barth, H. G.; Urban, M. W., Ed., Advances in Chemistry Series 247; American Chemical Society, Washington, DC, 1995; pp. 183-196.
14. van Asten, A. C.; Dam, R. J.; Kok, W. T.; Tijssen, R.; Poppe, H. *J. Chromatogr. A* **1995**, *703*, 245-263.
15. Gunderson, J. J.; Caldwell, K. D.; Giddings, J. C. *Sep. Sci. Technol.* **1994**, *19*, 667.
16. van Asten, A. C.; Boelens, H. F. M.; Kok, W. Th.; Poppe, H. *Sep. Sci. Technol.* **1994**, *29*, 513-533.
17. Martin, M.; van Batten, C.; Hoyos, M. *Anal. Chem.* **1997**, *69*, 1339-1346.
18. Rudin, A.; Johnston, H. K. *J. Polym. Sci., Polym. Phys. Ed.* **1971**, *9*, 55.
19. Rue, C. A.; Schimpf, M. E. *Anal. Chem.* **1994**, *66*, 4054-4062.
20. Schimpf, M. E.; Myers, M. N.; Giddings, J. C. *J. Appl. Polym. Sci.* **1987**, *31*, 117-135.
21. Jeon, S. J.; Lee, D. W. *J. Polym. Sci.: Polym. Phys. Ed.* **1995**, *33*, 411-416.
22. Cao, F.; Myers, M. N.; Giddings, J. C. *Anal. Chem.*, submitted.

Chapter 11

Colloidal Particles as Immunodiagnostics: Preparation and FFF Characterization

Teresa Basinska[1] and Karin D. Caldwell[2]

University of Utah, Department of Bioengineering, 20 S. 2030 E., RM 108, Salt Lake City, UT 84112–9450

The preparation and characterization of two types of particles with covalently attached proteins is described. The first type is a Pluronic F108 coated polystyrene latex (PS) where the surfactant has been end-group activated with pyridyl disulfide groups (PS-F108-PDS) for the subsequent protein attachment, while the second type is a synthesized poly(styrene/acrolein) latex (PSA), whose surface aldehyde groups allow a direct immobilization of protein via Schiff's base formation. Human immunoglobulin (IgG) was first covalently bound to the surface of these particles and in a second step a polyclonal antibody directed against this protein, namely a rabbit anti human IgG (α-IgG), was specifically adsorbed. Determinations of the surface concentrations of protein adsorbed in the two steps were accomplished by the sedimentation field-flow fractionation (SdFFF) technique, as well as by the more conventional bicinchoninic acid (μBCA) and amino acid analysis (AAA) methods, whereby a reasonable agreement was found between the three. It is shown that the mode of linking antigen (IgG) to surface has a significant influence on the efficiency of its antibody binding, with the tethered attachment providing greater access to the antibody. This study represents the first example of SdFFF characterization of a multi-layered composite where the different layers consisted of different chemical species.

In recent years, polymer particles (microspheres and latices) have found many applications in biochemistry and pharmaceutics. Particularly the fields of immunology and medical diagnostics have relied extensively on the use of colloidal particles in the

---

[1]Current address: Department of Polymer Chemistry, Center of Molecular and Macromolecular Studies, Polish Academy of Sciences, ul. Sienkiewicza 11290-363 Lodz, Poland.
[2]Current address: Center for Surface Biotechnology, Uppsala University, Box 577, S-751 23 Uppsala, Sweden.

form of latex medical diagnostic tests, where particle aggregation becomes a readily monitored measure of antigen/antibody concentration. For such tests to work well it is necessary to have either of the two, most commonly the antibody, immobilized at the surface of a latex particle without significant loss of activity. Properly immobilized, the antibody is then able to detect its soluble antigen and form clusters whose size and number reflect the level of presence of this antigen.

In the present work, two types of latices have been tested in a model study. The first type is a standard polystyrene latex with an adsorbed surface coating, consisting of the surfactant Pluronic F108. This surfactant is a triblock copolymer composed of two terminal polyethylene oxide (PEO) blocks attached to a polypropylene oxide (PPO) core. The PPO core strongly adsorbs to hydrophobic surfaces, such as polystyrene [1], while the flanking PEO chains, which are highly water soluble, extend out from the surface and form a barrier against nonspecific adsorption of proteins. A modification of the end hydroxyl group of PEO through the introduction of a pyridyldisulfide group has provided the possibility of covalently linking proteins with exposed sulfhydryl groups to the polymer. These reactive sulfhydryls may either be natural parts of the protein or may have been introduced through the reaction with a thiolating reagent, e.g. N-succinimidyl 3-(2-pyridyldithio)-propionate (SPDP), which takes place under mild conditions with respect to pH and ionic strength. Although this approach has many potential advantages, most notably the ease of functionalizing surfaces to a specified degree, it must be established to what extent the PEO barrier allows a large protein such as IgG to approach and become attached to the shielded surface. The second type of immunodiagnostic latex particles is synthesized in house from a 20:1 monomer mixture of styrene and acrolein. The synthesis protocol [2] results in a composite latex featuring a PS core with a thin acrolein shell, rich in reactive aldehyde groups to which IgG can be covalently bound via Schiff's base formation with free amino groups on the protein surface.

Here, we have immobilized the immunoglobulin G (IgG) molecule on the two particle types, and determined the resulting protein surface concentrations by a variety of different techniques. Specifically, this study focuses on establishing relationships between the extent of human IgG binding to these qualitatively different surfaces and the corresponding amounts of polyclonal antibody (rabbit anti human IgG) that can bind to each immobilized IgG molecule, i.e. the specific activity of the surface-bound immunoglobulin, in the two systems.

Previous studies [3-6] have demonstrated the SdFFF to be a technique capable of convenient and accurate assessments of the surface concentrations that result when various macromolecules adsorb to colloidal substrates of different size. In the present work, we utilized this technique to study the covalent binding of IgG on the surfaces of our different substrates.

Several analytical approaches can be devised to determine the surface concentration of materials adsorbed to nanoparticles of the type discussed here. These methods can be classified as being either "direct" or "indirect", where the latter are based on quantification of the losses of solute sustained by the supernatant following exposure to a particle sample with a known surface area. The "direct" methods, in turn, evaluate the amounts of adsorbate directly associated with a given amount of particles. Both approaches are fraught with technique-specific sources of error, such as those associated with the ability to accurately determine the particle surface area available for adsorption in a given experiment. In addition, the "indirect" methods will include in the estimated protein surface concentration not only the firmly adsorbed material, but loosely adherent structures as well. The "direct" methods, in turn, require analytical techniques capable of accurately quantifying the ad-layer in the presence of the particulates - a requirement that removes many optical techniques from consideration unless a quantitative removal of particles can be accomplished prior to the analysis.

The previously referenced articles detail an approach to sedimentation field-flow fractionation (SdFFF) that allows a "direct" determination of the mass increase per particle from observed differences in retention between the bare and coated particles [4]. This technique has the advantage over other "direct" methods in that it determines the mass uptake per particle, without any form of labeling, in a manner that leaves the particles well washed and free from loosely adherent material.

The SdFFF technique is an analytical separation method that relies on the coupled influences of an applied sedimentation field and a perpendicular flow of sample through a thin channel. The technique allows an exact determination of particle mass from the observed level of retention of a particulate sample under a given field, provided the densities of particle and suspension medium are known. If the particle is allowed to adsorb or covalently bind a coating of known unsolvated density, the added mass translates into an increased retention under the chosen field. Through its ability to produce measurable retention shifts for coated samples, the sedFFF acts as a very sensitive microbalance. Since the surface area per particle is directly determined from the retention of bare particles under a given field strength, the mass per unit area is readily calculated from these measurements.

Materials

Latexes. Polystyrene latex (PS) standard particles (Bangs Laboratories, Inc.) with nominal diameter 150 nm were used as substrates in the adsorption experiments, while the adsorbing compound was a derivatized Pluronic F108 (BASF Corporation, molecular weight 14,600) equipped with pyridyldisulfide groups (F108-PDS). The method of derivatization of Pluronic F108 was described earlier [7]. The unsolvated density of this surfactant had been determined earlier to be 1.16 g/mL.

During the adsorption experiment, the PS particles (1% suspension) were incubated with F108-PDS (total volume 1 mL and concentration 4% (w/w)) in phosphate buffered saline (PBS, I=0.2M, pH=7.4, density 0.997 g/mL at 25 °C) for 16 hours at room temperature under constant end-over-end shaking. After the adsorption process, the coated latex particles were centrifuged for 30 min. at 30,000 g (Beckman, Optima TL Ultracentrifuge), whereupon a series of 800 µL aliquots of the supernatant were removed and replaced with buffer solution. This procedure was repeated 6 times to remove essentially all unadsorbed surfactant.

A batch of poly(styrene/acrolein) latex (PSA) was prepared through radical polymerization of styrene and acrolein, which was initiated with $K_2S_2O_8$, in water; the reaction was carried out at 80 °C without emulsifier, and was allowed to proceed for 28 hours. The mole fraction of acrolein in the monomer mixture was 0.050. Details of the synthetic procedure, and the methods used for latex characterization, were published earlier [2].

Proteins. The protein bound to the latex surfaces was a human IgG (Sigma Immunochemicals), serving as antigen, and a polyclonal anti human IgG (Gamma-Chains) rabbit antibody (DAKO A/S, Glostrup, Denmark), serving as antibody. The densities of both types of IgG were assumed to be 1.353 g/mL. For studies of the antigen binding activity displayed by the antibody we used anti human IgG rabbit antibody conjugated with alkaline phosphatase (Sigma). N-succinimidyl 3-(2-pyridyldithio)-propionate, SPDP (Pierce) was used for thiolation of human IgG. The protein used for non-specific adsorption onto the IgG coated PSA particles was human serum albumin (HSA) (Sigma).

Techniques and Methods

Characterization of Bare and Coated Particles. The bare and Pluronic coated polystyrene particles (PS) and the bare and protein coated poly(styrene/acrolein) particles (PSA) were sized by means of photon correlation spectroscopy using a Brookhaven Model BI-200 variable angle instrument with a BI-2030 autocorrelator; the light source for this instrument was 30 mW HeNe laser (Spectra Physics) emitting at 732 nm. The concentration of functional groups at the surface of the PSA particles was measured in a reaction with 2,4-dinitrophenylhydrazine (DNPH), which is an analytical reagent for determination of aldehyde groups [8-9]. Details concerning this determination were described earlier [10].

The surface concentration of pyridyl disulfide groups tethered to the Pluronic coated PS particles was determined using a method described by Carlsson at al. [11]. Briefly, a known amount of Pluronic F108-PDS coated PS particles (PS-F108-PDS) in PBS buffer (pH=7.4) was incubated with 0.1 mL 25 mM DTT in an Eppendorf tube, for 10 min. Next, this sample was centrifuged at 16,000 g for 30 min. to spin down the particles. By spectrophotometric measurement of released 2-thiopyridone, the amount of pyridyl disulfide groups was determined, which in turn allowed a determination of the amount of F108-PDS adsorbed to these particles, and thereby the amount of available protein linking sites.

Derivatization of IgG. In order to bind IgG to the activated ends of the Pluronic PEO blocks, it was necessary to first introduce free sulfhydryl groups into the IgG macromolecule. This was accomplished through a reaction with the bifunctional reagent N-succinimidyl 3-(2-pyridyldithio)-propionate (SPDP) [12]. To 5 mg of IgG in 1 mL of PBS buffer (pH=7.4) was added a 25 μL aliquot of 20 mM stock solution of SPDP in DMSO. The sample was incubated for 30 min. at room temperature. The excess of SPDP was then removed by gel permeation chromatography using a PD-10 column (Pharmacia), and the fractions (1 mL per fraction) containing protein were collected. The pyridyldisulfide groups introduced into the IgG macromolecule were then cleaved by dithiothreitol (DTT, Aldrich). The ratio of IgG to DTT was kept at 3.76:1 (w/w). To the protein in PBS solution was added 3.5 mM ethylenediaminetetraacetate (EDTA) in 0.2 M Tris buffer (pH=8.1) in order to chelate any contaminating metal ions. The reduction reaction was then carried out at room temperature for 45 min. Excess DTT was again separated from protein by a PD-10 column, equilibrated in 0.1 M sodium phosphate buffer (pH=6.0) containing 5 mM EDTA. After this separation, the fraction containing thiolated IgG was immediately added to a suspension of Pluronic F108-PDS coated PS particles.

Human Thiolated IgG Coupling to Pluronic F108-PDS coated PS Particles. A portion of thiolated human IgG (600 μL of 3.5×10^{-4} g/mL) in phosphate buffer (pH=6.0) was incubated with F108-PDS coated PS particles (400 μL of 3×10^{-3} g/mL) in an Eppendorf tube for four hours on a rotating wheel with constant end-over-end shaking at room temperature. After this coupling reaction (see Scheme 1) the coated latex particles were centrifuged for 15 min. at 20,000 g (Beckman, Optima TL Ultracentrifuge), whereupon a series of 800 μL aliquots of the supernatant were removed and replaced with buffer solution. This procedure was repeated 5 times. The first supernatant was left for determination of protein by the micro bicinchoninic acid method (μBCA, kit purchased from Pierce). After the first washing, 5 μL of sample was taken for analysis in the SdFFF system. The 200 μL portion of IgG -bound PS-F108 particles was left for amino acid analysis (AAA).

Scheme 1.

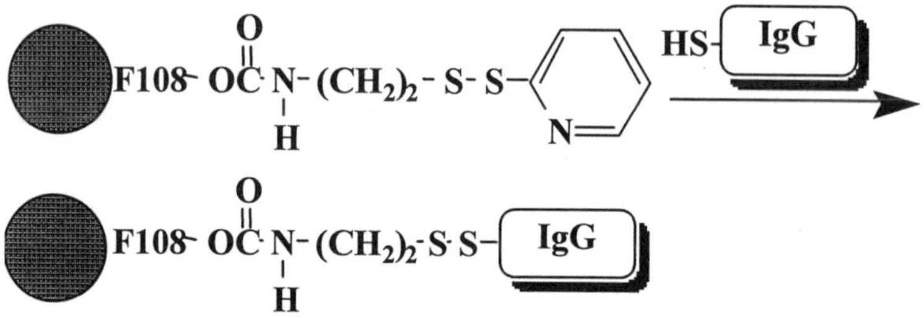

Human IgG Coupling to PSA Particles. In the case of PSA particles, 3.2 mL of a 7.8×10^{-4} g/mL IgG solution in Tris buffer (pH=8.3) was added directly to 6.8 mL of a latex suspension containing solids with a concentration of 3×10^{-3} g/mL. This sample was incubated for 20 hrs at room temperature, and after that time the same analysis procedure was applied as was outlined above for the IgG coated PS-F108 particles. The chemical reaction of human IgG coupling to PSA particles is presented in Scheme 2 below.

Formation of Antigen-Antibody Complex. The 40 μL sample of PS-F108 particles with covalently bound human IgG was exposed to rabbit anti human IgG (5 μL) for 15 minutes at room temperature. Usually, an 8 μL aliquot of this reaction mixture was taken for injection into the SdFFF channel. The excess of antibody was removed from the remainder by centrifugation. This procedure was repeated four times, until the

Scheme 2.

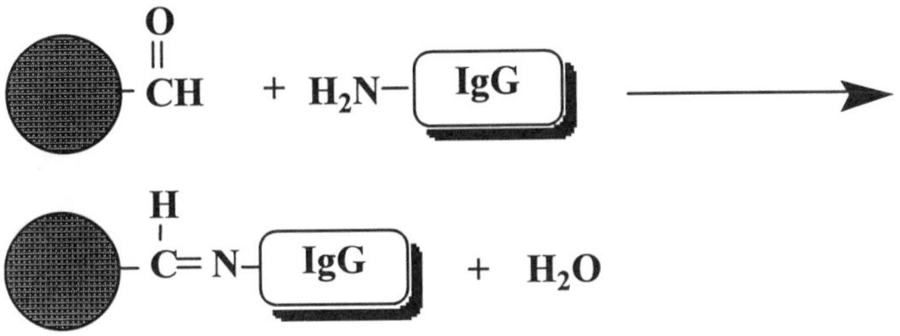

$$\text{(particle)}-\overset{O}{\overset{\|}{C}}H \quad + \quad H_2N-\boxed{\text{IgG}} \quad \longrightarrow$$

$$\text{(particle)}-\overset{H}{\overset{|}{C}}=N-\boxed{\text{IgG}} \quad + \quad H_2O$$

supernatant was free from protein. This particulate sample was left for amino acid analysis (AAA).

Assay Method for Determination of Efficiency of Antibody Binding (a-IgG). The efficiency of antibody binding was determined using an alkaline phosphatase-conjugated rabbit anti-human IgG (Sigma), essentially as described by Bretaudiere and coworkers [13]. This assay consists of a one step reaction involving the hydrolysis of 4-nitrophenyl phosphate (disodium salt), which gives rise to the intensely yellow-colored reaction product nitrophenol that can easily be quantified spectrophotometrically through its absorbance at 405 nm (see Scheme 3).

Scheme 3.

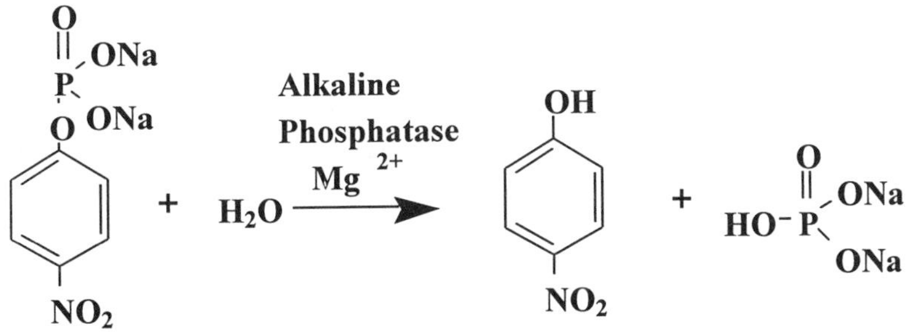

$$\text{4-nitrophenyl phosphate} \quad + \quad H_2O \quad \xrightarrow[\text{Mg}^{2+}]{\text{Alkaline Phosphatase}} \quad \text{nitrophenol} \quad + \quad HO-\overset{O}{\overset{\|}{P}}\overset{ONa}{\underset{ONa}{}}$$

To 600 μL samples of human IgG, coated on either type of particles (taken from respective stock suspension) 20 μL aliquots of AP-conjugated anti human IgG were added. The samples were incubated for 40 min. at room temperature under constant shaking, and unbound conjugate was subsequently removed by centrifugation. Sample aliquots of 20 μL were then taken for measurements of efficiency of antibody binding, leaving 600 μL samples for determination of bound protein by amino acid analysis.

For measurements of efficiency of antibody binding we used 0.3 mL of solutions containing 1 mg/mL of 4-nitrophenyl phosphate in a 0.1 M diethanolamine-HCl buffer (pH=9.8). This substrate was usually added to 0.2 mL of alkaline phosphatase conjugate, either free or specifically bound to human IgG immobilized on particles, and the volume of the mixture was adjusted to 1 mL through the addition of diethanolamine-HCl buffer. Next, the sample was incubated at 30 °C, for 30 min. Exactly after that time the optical density at 405 nm was registered.

In each experiment we used known amounts of latex particles with well determined average particle diameters and, hence, surface areas. All spectroscopic measurements were made using a Perkin-Elmer Lambda 6 UV/VIS Spectrophotometer.

Theory of SdFFF

The theoretical basis for deriving sample size/mass information from retention measurements made with sedimentation FFF has been discussed in detail elsewhere [4, 14,15]. For the purpose of this communication we will recapitulate the basic equations needed to convert observed retention volumes into desired characteristics of the coated particles. In the present work, these characteristics are the size of the bare particles and the mass of the adsorbed (in the case of Pluronic or HSA), or covalently attached layer (in the case of IgG and α-IgG), from which one derives a value for the protein surface concentration.

A sample introduced into the thin FFF channel is forced by the applied sedimentation field to migrate towards one of the walls perpendicular to the field, where it concentrates. As a result of the interplay between diffusion and field-induced concentration the sample will equilibrate into an exponentially distributed zone whose highest concentration is found at the wall. Thus, the concentration along the field axis is described by

$$c(x) = c(0) \exp(-x/\lambda w) \tag{1}$$

where $c(0)$ represents the concentration at the accumulation wall ($x=0$), and w is the distance between the major walls of the thin channel. Parameter λ represents the dimensionless thickness of the solute cloud, i.e. essentially its average extension into the channel divided by w. Under the influence of a sedimentation field with acceleration G, a particle of mass m_1 will experience a force F_1 equal to

$$F_1 = m_1 (1 - \rho_3/\rho_1) G \tag{2}$$

In this expression the parenthesis reflects the buoyancy factor, with ρ_1 and ρ_3 representing the densities of the particle and carrier medium, respectively. In case the particle is spherical with diameter d_1, mass m_1 in the above expression can be replaced by the product of volume and density to yield the following modified force expression

$$F_1 = (d_1^3 \pi/6) \Delta\rho G \tag{2a}$$

where $\Delta\rho$ represents the density difference between particle and carrier. In its general form, valid for all types of FFF, parameter λ expresses the ratio of a sample's thermal energy and its average energy in the presence of the applied field [16]

$$\lambda = kT / Fw \tag{3}$$

For the specific case of a sedimentation field, force F is given by Eqs. (2), or (2a), and the characteristic concentration distribution will therefore be governed by a λ expression of the following form

$$\lambda_1 = kT \, / \, m_1 \, (1-\rho_3/\rho_1) \, Gw = 6kT \, / \, d_1^3 \, \pi \, \Delta\rho \, Gw \qquad (4)$$

Immediately following an injection of sample into the channel, the flow of mobile phase is stopped while each sample component relaxes into its specific equilibrium distribution, described by Eq. (1). At the end of this relaxation period, a complex sample is forming a zone at the head of the channel, which is a composite of the exponential distributions generated by each component. As the channel flow is initiated and allowed to establish its parabolic distribution of velocities perpendicular to the field, each distribution cloud will be transported downstream in accordance with its extension into the velocity field. The retention ratio R, defined as the ratio of zonal and mobile phase velocities, can therefore be expressed in a general form, valid for all types of FFF [16]

$$R = V^o/V_r = 6\lambda \, [coth(1/2\lambda) - 2\lambda] \qquad (5)$$

The experimentally observed parameters V^o, the column void volume, and V_r, the retention volume, are the basis for determining an empirical λ-value which is convertible into a particle mass or diameter through use of Eq.(4).

Evaluation of particle size from retention data requires knowledge of the densities of both particle and mobile phase, as evident from Eq.(4). In general such data are readily available as long as the particle remains of uniform composition. The situation changes drastically when the particle is allowed to adsorb material of a density different from its own, and the retention will therefore reflect the composite mass of core particle and adlayer. Assuming the density of the adsorbing species to be ρ_2, and the mass adsorbed (in the case of Pluronic F108 or covalently bound protein in the case of IgG) per particle to be m_2, one finds parameter λ for the composite to be [4, 5]

$$\lambda_{comp} = kT \, /\{[m_2(1-\rho_3/\rho_2) + m_1(1- \rho_3/\rho_1)]Gw\} \qquad (6)$$

For analyses of multi-layered shells around a given core particle, equation 6 can be expressed in the more general form:

$$\lambda_{comp} = kT/ \, Gw[\textstyle\sum_i m_i(1\rho_s/\rho_i)] \qquad (7)$$

where ρ_s symbolizes the density of the carrier solution and subscript i in the sum is allowed to run from 1 (the core particle) to however many layers the analysis will cover.

In the simple case of one ad-layer with mass m_2, it is helpful to reorganize eq.6 by substituting the product of volume and density for the bare particle for m_1, i.e.

$m_1 = \pi d_1^3 x \, \rho_1/6$. In this way it is possible to express adsorbed mass m_2 in terms of the retention parameter λ_{comp} determined for the coated particles under a given sedimentation field G

$$m_2 = [kT/(Gw\lambda_{comp}) - \pi d_1^3 \, (\rho_1-\rho_3)/6] \, (1-\rho_3/\rho_2)^{-1} \qquad (8)$$

Since the area A per bare particle is πd_1^2, the mass per unit area, or surface concentration Γ, is directly obtainable from retention measurements in SdFFF [17]

$$\Gamma = m_2/A = [kT/(Gw\lambda_{comp} \pi d_1^2) - d_1(\rho_1-\rho_3)/6] (1-\rho_3/\rho_2)^{-1} \qquad (9)$$

Practical Considerations Regarding the Sd FFF Analysis

All measurements were done using a SdFFF instrument built in-house and connected to a computer, which controls the spin rate and fluid delivery routine and collects data from the detector and effluent-monitoring balance as a function of time. Inside the rotor (radius 15.5 cm) is a flow channel with a thickness of 0.0254 cm and a length of 96 cm from tip to tip of the V-shaped in/outlets. The channel breadth is 2.0 cm and the void volume (V^o) equals 4.80 mL. This value was determined by injections of acetone, an unretained sample. Elution volumes were measured as weights on a continuously recording electronic balance connected to the computer.

Samples, usually 3-10 µL were injected through a septum directly into the channel. After pumping 0.4 mL of carrier into the system, the pump was stopped and the system was left to relax under the chosen RPM. At the end of the chosen relaxation time, the pump began its delivery of carrier to the channel.

Results and Discussion

The purpose of this study was to compare two modes of immunoglobulin attachment to particulate carriers, in terms of the resulting protein surface concentrations, and in terms of the subsequent access to the attached protein by a specifically binding macromolecule. Of particular interest was to determine the degree to which large proteins, such as the IgG molecule, can approach and bind to a carpet of PEO tethers present on the particle surface. PEO-attachment to surfaces is usually accomplished to prevent protein access, and it might well be argued that attempts to attach immunoglobulins to end-group activated Pluronic F108 might be unproductive.

Basic Characteristics of the Commercial Polystyrene Latex (PS). The polystyrene latex used here had a nominal diameter of 150 nm. Its diameter measured by PCS was equal to 152 nm, while a value of 151 nm was provided by sedFFF (see Table I). The Pluronic F108-PDS was adsorbed to the surface of these PS particles with a concentration of PDS groups determined to be 1.62×10^{-8} mol/m^2. This determination was performed spectroscopically, following reductive cleavage of the disulphide bridge and quantification of the released thiopyridone.

Basic Characteristics of the Synthesized Poly(styrene/acrolein) Particles (PSA). In previous work by one of us [2] it was established that the PSA latices produced according to the method described above have core-shell structures, with the core rich in polystyrene and the shell rich in polyacrolein. In the present work, the size of the produced PSA particles was determined by PCS to be 372 nm, with a specific polydispersity index of 0.007. This small number indicates that these particles were virtually monodisperse. We also determined the concentration of acrolein derived aldehyde groups at the surface and found this number to be 1.90×10^{-6} mol/m^2.

Study of bare and coated particles by sedFFF. The ability to measure the mass uptake by colloidal particles through measurements of retention changes in SdFFF was addressed in the introduction. Here, we compare the results of two modes of attaching proteins to particles, namely: 1. through a direct, covalent linkage to the PSA particles, and 2. through a two-step linking process that involved the adsorption of F108-PDS to PS latex as a first step, and the subsequent replacement of the PDS group by a thiolated protein as the second step.

Figures 1 and 2 display fractograms obtained after injections of aliquots of bare and coated (F108 and/or protein) particles into the SdFFF system. The symmetrical peaks indicate that the injected particles represent uniform distributions, free from aggregates. The singlet peak retention volumes were determined with considerable precision, as judged from three analyses of each sample, keeping the experimental conditions constant. With variations in retention volume being no more than ±0.6 mL, the measurements yielded a standard deviation in the determined protein uptake of 0.1 mg/m^2. Tables I and II summarize mass/size determinations for the bare particles and their coating layers. The third column of each table lists the sizes of the bare particles, determined from the positions of the peak maximum during the SdFFF measurement, as well the sizes of particles (bare and coated) from determinations by PCS.

The PCS-derived thickness of the Pluronic coating on the PS particles (Table I) is equal to 17 nm, and the Pluronic mass uptake per PS particle is 1.96×10^{-16} g. These numbers are in good agreement with data reported previously from this laboratory [4].

Table I. Parameters characterizing the bare and coated polystyrene particles (PS)

Particles [nm]	Coating type	Size measurement [nm]	V_e [mL]	Total mass uptake[g]
PS 150	bare	151± 1(SdFFF) 152± 1(PCS)	29.20 ±0.2	19.10×10^{-16} *
PS 150	F108-PDS	186± 7 (PCS)	37.90±0.3	1.96×10^{-16}
PS 150	F108-PDS+IgG	193± 6 (PCS)	49.50 ±0.7	3.61×10^{-16}
PS 150	F108-PDS+IgG+α-IgG	-	-	7.43×10^{-16} **

* mass of the bare particle, as determined by sedimentation FFF
** mass uptake, determined by AAA

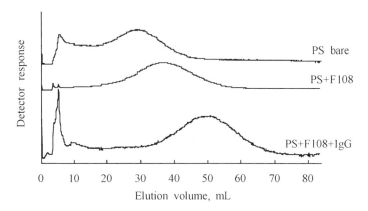

Figure 1. SdFFF fractograms of bare and coated PS particles. Experimental conditions: Field: 2000 rpm (694g); stop flow time: 20 min.; carrier: 5 mM glycine buffer, pH 8.3; flow rate: 2.3 mL/min.

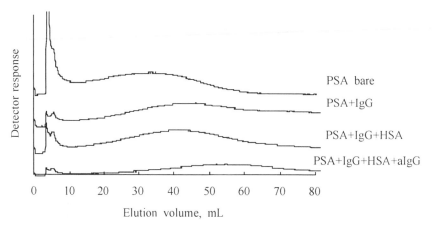

Figure 2. SdFFF fractograms of bare and coated SPA particles. Experimental conditions: field: 500 rpm (43g); stop flow time: 30 min.; carrier: 5 mM glycine buffer, pH 8.3; flow rate: 2.3 mL/min.

The elution volume (V_e) corresponds to the position of the peak maximum for each collected fractogram. The equations presented in the theory section outlined the procedure for calculating the mass of each particle and the uptake of protein by each single particle. For both types of particles the mass uptake in each step increased significantly, as demonstrated by an increased elution volume in the fractogram. In case of the poly(styrene/acrolein) particles with bound IgG, human serum albumin (HSA) was applied in a second coating step in order to cover up any unsaturated

Table II. Parameters characterizing the bare and protein coated poly(styrene/acrolein) latex (PSA).

Particles [nm]	Coating type	Size measurement [nm]	V_e [mL]	Total mass uptake[g]
PSA 372	bare	361±1 (SdFFF) 372±1 (PCS)	32.6±0.5	25.06x10^{-15} *
PSA 372	IgG	383±4 (PCS)	43.0±0.4	2.18x10^{-15}
PSA 372	IgG+HSA	384±3 (PCS)	43.3±0.2	2.23x10^{-15}
PSA 372	IgG+HSA+α-IgG	395±7 (PCS)	54.2±0.2	4.20x10^{-15}

* mass of the bare particle, as determined by sedimentation FFF

particle surface to prevent nonspecific binding of the anti-IgG (α-IgG), to be added in the following step. Only insignificant amounts of HSA appeared to adhere to the IgG coated particles, as demonstrated by the lack of a peak shift associated with this step (see Fig. 2), and thus essentially the entire surface appeared to be saturated by IgG in the previous step.

In addition to providing analytical information about the particles and their mass uptakes, the SdFFF process yielded stable fractions that could be further analyzed by PCS to determine the thickness of the bound layers. It is important to stress that, although the SdFFF allows a determination of the size of bare particles, i.e. particles of uniform composition, the technique is unable to provide a size for composite particles, of the type produced through the various coating reactions described here. This is due to the unknown arrangement and degree of solvation of the molecules in the ad-layer. However, the fact that eluate of uniform composition can be collected and sized by an independent technique, such as PCS, is a clear advantage of the SdFFF method.

As shown in Table 1, there is a mass uptake of Pluronic F108-PDS on the PS latex amounting to 1.96x10^{-16} g per particle. From the determined number of PDS groups per F108 molecule, we calculate the surface concentration of active linking sites to be 1.62x10^{-8} mol/m^2. It is therefore interesting to note that the 3.61x10^{-16} g of

IgG bound per particle via the F108 linker translates into a surface concentration of 1.48×10^{-8} mol/m^2, or very nearly the same number as the number of available end-groups. We can therefore conclude that the attachment to the PEO chains of even such large proteins as the IgG can occur without notable steric hindrance.

Comparison of Methods for Determination of Protein Surface Concentration. The surface concentrations of bound protein were determined by two methods in addition to the SdFFF. One of these was the "direct" method of amino acid analysis (AAA), whereby the protein associated with a given amount of particles was hydrolyzed, and the generated amino acid residues quantified. This method has been used extensively in our laboratory in conjunction with studies of the fouling of biomaterials, such as contact lenses. We estimate the errors in the AAA analysis of protein in the presence of a polymer matrix to amount to about 10%. The sedFFF data were also compared with results from an "indirect" method, namely the microbicinchoninic method (μBCA), which presents uncertainties in the colorimetric protein determination of about 5%. In addition to errors in the actual protein determinations, this technique measures the amounts of immobilized protein as differences in concentration before and after binding, which often means the small difference between two big numbers. Together with general errors introduced in the assessments of surface area available for adsorption, the indirect method also includes errors due to the amounts of protein that is loosely associated with the particle, as discussed in the Introduction. As mentioned previously, none of these errors affect the SdFFF measurement, since with this technique the mass uptake is measured on a *per particle* basis, with the area of the core particle being well established.

The results of this comparison are presented in Table III. Here, numbers obtained from SdFFF measurements (in grams per particle, see Tables I and II) have been converted into mg per square meter of particle surface, in conformity with the numbers obtained from μBCA and AAA. Overall, there is good agreement between the SdFFF determinations and those provided by the other techniques, in that all three indicate a density of coverage for IgG on the PSA particles which is twice as high as that found on the Pluronic F108-PDS coated particles.

Table III. Comparison of methods for determining protein surface concentration.

Type of coating	Bound protein (mg/m^2) determined by:		
	SdFFF	μBCA	AAA
PSA+IgG	5.5	3.25	3.3
PSA+IgG+HSA	5.6	-	3.9
PSA+IgG+HSA+a-IgG	10.6	-	9.1
PS-F108-PDS+IgG	2.3	2.4	1.7
PS-F108-PDS+IgG+α-IgG	-	-	5.3

As concerns the specific binding of antibody to the IgG-antigen on the particle surface, this second layer of protein was readily quantified for the PSA particles, but not for the PS counterparts. Unfortunately, the latter formed large amounts of aggregates upon antibody addition, and could not be eluted from the SdFFF channel. For these particles, the protein uptake in the specific binding step was therefore only determined through AAA, as indicated in Tables I and III.

Effect of Attachment Mode on the Efficiency of Antibody Binding After measuring the protein uptakes by the two types of particles, it was possible to calculate a ratio between the surface bound antigen (human IgG) and its specifically bound antibody (rabbit anti human IgG) for the two. For the PSA, these calculations were based both on data obtained from sedFFF and from amino acid analysis; the two gave the same value of 1.6. For Pluronic-PDS coated particles the obtained ratio was calculated from the AAA values only, and gave a value of 2.2. The difference between these numbers indicates a more favorable presentation of the antigen in the case of the tethered attachment, which apparently allows a more effective interaction with the specifically binding antibody.

Efficiency of the AP conjugate. In this work, the chosen assay for anti IgG is based on the processing of the low molecular weight substrate p-nitrophenyl phosphate by the alkaline phosphatase attached to the antibody. It is reasonable to believe that the mode of attachment of the antibody to its antigen would reflect the rate with which the substrate is turning over into the colored product, p-nitrophenol. Thus, a given amount of AP should show the highest activity when free in solution, and the lowest when the protein is attached very near a surface, where mass transfer problems become severe. The tethered arrangement could be thought to display an intermediate activity due to facilitated access to the enzyme. From the assays compiled in Figure 3, it is clear that the expected behavior is indeed taking place. Specifically, the relationships between the amounts of released nitrophenol (measured at 405 nm) and the concentration of AP-conjugate are linear. The highest slope was recorded for the free conjugate in solution, the intermediate slope was found for the anti-IgG conjugated with AP and bound to IgG immobilized on PS-F108-PDS latex, and the weakest slope was determined for the conjugate associated with the PSA particles.

Conclusions.

The present work compares two modes of attachment of the IgG molecule to latex particles with potential use as immunodiagnostic reagents. The two modes involve either direct binding to the particle surface, or binding via a PEO tether whose molar mass is around 5000 Da; this tether is part of a tri-block copolymer of the PEO-PPO-PEO general structure that has a molar mass of 14,600, as reported by the manufacturer. Although the direct binding gave the highest surface concentration, the protein in this configuration did not bind as much of its specific affinity ligand as it did in the tethered arrangement. Furthermore, the tether appeared to facilitate access to the surface-bound protein by low molar mass components compared to the situation for the directly attached protein on the PSA particles. It is clear from this work that the PEO tethers do not prohibit the covalent attachment of large molecules, such as the

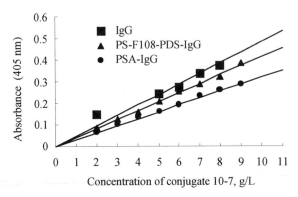

Figure 3. Phosphatase activity for different forms of conjugate complex. The upper trace represents a complex with soluble IgG, while the two lower traces reflect complexes with particle-bound IgG. The abscissa represents concentrations of AP-conjugated anti-IgG.

IgG, to the active leaving groups at the end of the PEO blocks, as has often been suggested. Instead, there appears to be a virtually complete replacement of existing PDS groups with thiolated IgG molecules on our Pluronic coated latex particles.

Three different analytical techniques were used in this work, and their results are shown to be in reasonably good agreement with one another. Among them, the SdFFF method proved particularly attractive because of its high reproducibility. Other advantages include its ability to directly produce values for the mass uptake resulting from the different reaction steps to which the latex particles were exposed, without the need for error prone estimates of the particle surface area available for adsorption, and without the need for cumbersome wash procedures. Such needs are clearly obviated by the wash accomplished by the fractionation process itself.

The present study represents the first SdFFF analysis of a multistep adsorption process in which particles are successively taking up multiple components of different density (surfactant and protein, respectively).

References.
1.   Li J.; Carlsson J.; Huang S.C.; Caldwell K.D. *Hydrophilic Polymers; Performance with Environmental Acceptability*, ACS ser. 248, 1996;p.p.62-78
2.   Basinska T.; Slomkowski S.; Delamar M. *J. Bioact. Compat. Polym.* **1993**, *8*, 205
3.   Caldwell K.D.; Li J.T; Dalgleish D.G. *J. Chromatogr.* **1992**, *604*, 63
4.   Li J.T.; Caldwell K.D. *Langmuir* **1991**, *7*, 2034
5.   Beckett., R.; Ho J.; Jiang Y.; Giddings J.C. *Langmuir* **1991**, *7*, 2040
6.   Langwost B.; Caldwell K.D. *Chromatographia* **1992**, *34*, 317
7.   Li J-T.; Carlsson J.; Lin J-N.; Caldwell K.D. *Bioconjugate Chem.* **1996**, *7*, 592
8.   Cheronis N.D.; Ma T.S. *Organic Functional Group Analysis by Micro and Semimicro Methods*, Interscience, New York, US, 1964, p.145
9.   Peters T. *Adv. Protein Chem.* **1985**, *37*, 161
10. Slomkowski S.; Basinska T.*Polymer Latexes: Preparation, Characterization and Applications*; Daniels E.S., Sudol E.D., El-Aasser M.S. Eds., ACS Symp. Ser. 1992, *492*, p.p.328-339
11. Carlsson J.; Drevin H.; Axén R. *Biochem. J.* **1978**, *173*, 723
12. Neurath A.R.; Strick N. *J. Virol. Meth.* **1981**, *3*, 155
13. Bretaudiere J.P.; Spillman T. *Methods of Enzymatic Analysis*, Bergmeyer H.U. Ed.; Vol.IV, Verlag Chemie, Weinheim, DE, 1984, p.p. 75-82
14. Caldwell K.D. *Modern Methods of Particle Size Analysis* H. Barth Ed.; John Willey & Sons, New York, US, 1984, p. 220-232
15. Karaiskakis G.; Myers M.N.; Caldwell K.D.; Giddings J.C. *Anal. Chem.* **1981**, *53*, 1314
16. Giddings J.C.; Caldwell K.D. *Physical Methods of Chemistry*, Rossiter B.W., Hamilton J.F., Eds., John Willey & Sons, New York, US, 1986, Vol. IIIB; p.867
17. Caldwell K.D.; Li J.-M.; Li J.-T.; Dalgleish D.G.*J. Chromatogr.* **1992**, *604*, 63

Chapter 12

Molecular Characterization of Complex Polymers by Coupled Liquid Chromatographic Procedures

Dušan Berek

Polymer Institute of the Slovak Academy of Sciences, 842 36 Bratislava, Slovakia

Selected „coupled" liquid chromatographic (LC) methods are briefly reviewed, that are aimed at separation and molecular characterization of polymer systems composed of chemically different units. The latter systems are complex in their chemical and often also in their physical structure and exhibit multiple distributions. Consequently, size exclusion chromatography produces only semiquantitative data on their molecular characteristics. The coupled LC methods combine at least two separation mechanisms, typically entropy based size exclusion mechanism with one or several enthalpy based interactive mechanisms (e.g. adsorption, partition, solubility). The adsorption mechanism is presently the partner of choice for coupling with exclusion, though further interaction based mechanisms can also be employed. So far, published data indicate great potential of coupled liquid chromatographic procedures and it is anticipated that the latter will undergo a vivid development in the near future, especially in connection with necessity to characterize new specialty polymers.

High-performance polymeric materials usually exhibit a complicated or at least a non-uniform chemical structure. We speak about complex polymer systems such as polymer blends, copolymers, sequenced and functionalized polymers. Precise molecular characterization of these materials is very exacting and usually requires separation of polymer species differing in their molar mass, chemical composition, structure etc.

Presently, the most important method for analytical and preparative fractionation of macromolecules is size exclusion chromatography (SEC), also called gel permeation chromatography in the case of lyophilic synthetic polymers and gel filtration chromatography in the case of hydrophilic macromolecules, especially biopolymers. SEC employs the entropy based separation mechanism that discriminates polymers and oligomers according to their size in solution. SEC is known for its high precision and reliability, repeatability and speed, and further for its good reproducibility and low sample consumption. At the same time, however, the method exhibits several inherent drawbacks. For example, SEC suffers from limited both separation selectivity and resolution power. From the point of view of complex polymer systems, it is important that SEC cannot resolve macromolecules of the same size that may differ in their molar mass and, simultaneously in their composition. In other words, size exclusion chromatography cannot alone reliably discriminate the constituents of complex polymer systems.

Appropriate possibilities for both enhancing SEC separation selectivity and characterizing complex polymer systems offer liquid chromatographic (LC) methods employing combinations of different separation mechanisms, so called coupled LC procedures.

The simplest couplings manage with one single column packing and with the isocratic elution mode. More sophisticated coupled LC methods apply column and/or eluent switching. Alternatively, temperature or pressure in the chromatographic system can be varied in a controlled way. A single mechanism or coupled LC system that separates macromolecules on the base of exclusion or interactive or combined separation mechanisms can be connected - usually on-line - with further, independent LC system(s) again employing a single or combined separation mechanism. We speak about higher degree of LC coupling or about two- or multi-dimensional liquid chromatographic separations.

In this paper, we shall briefly discuss selected liquid chromatographic procedures combining the size exclusion separation mechanism with particular interactive separation mechanisms based on adsorption, partition and solubility. We shall limit our discussion to „first degree LC couplings" and only brief hints will be made at the potential candidates for LC couplings of a higher degree.

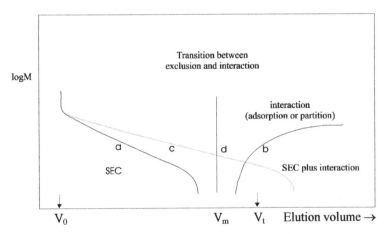

Figure 1 Liquid chromatographic calibration curves for macromolecules. V_o is interstitial volume, V_m is volume of liquid within column, V_t is total volume of column. For further explanation see text.

The basic terms we shall deal with are evident from a dependence of the logaritm of the molar mass of macromolecules (M) on the LC retention volume (V_R). This dependence is called a **calibration curve** in SEC and we shall use this term also in the cases where actually no calibration is considered. Schematic calibration curves for exclusion (a) and interaction separation (b) mechanisms are shown in Figure 1. The couplings that lead either to the separation selectivity rise (c) or to the transition situation, i.e. to the total loss of separation selectivity (d) in terms of the chain size are also featured.

Coupling exclusion with adsorption to enhance selectivity of SEC separation. Isocratic elution with a single column packing

As evident from Figure 1, the retention volumes decrease with rising molar mass of particular polymer chains in the SEC mode (curve **a**) while they increase with M in the liquid adsorption chromatographic mode (LAC, curve **b**). In order to enhance the SEC separation selectivity, a chromatographic system column packing/eluent must be identified in which only the end groups of macromolecules are adsorbed. In this case the effect of adsorption rises with decreasing polymer molar mass and results in enhanced retention volumes of smaller molecules. A practical example of this approach is represented by the separation of poly-ethyleneglycols (PEGs) using Sephadex LH-20 gel with various eluents (1). Highly polar eluents, e.g. methanol or dimethylformamide, suppress interactions between PEG and gel and a nearly "ideal" SEC separation is observed (curve **a** in Figure 1). On the other hand, less polar eluents such as tetrahydrofuran (THF) or acetone strongly promote adsorption of the -OH end groups of PEG. Consequently, the slope of the SEC calibration curve decreases, i.e. the separation selectivity increases (curve **c** in Figure 1). This approach can be used also for discrimination and simultaneous molecular characterization of oligomers with different functionalities, e.g. with different nature and/or number of their end groups. For example, ethoxylated nonylphenol was discriminated from the reaction by-product, poly-ethylene glycol, and both oligomers were simultaneously separated according to their molar masses (2). In this case a strong adsorption of phenyl groups on Sephadex LH-20 gel in methanol eluent was utilized. Unfortunately, a wider application of above approach is limited due to the lack of appropriate column packings.

A small SEC selectivity enhancement can be observed also with high polymers if eluent slightly promotes their adsorptive retention. Typically, polystyrenes eluted from silica gel based column packings exhibit more flat *universal* calibration dependence in toluene than in THF (3,4). As known, toluene is a weak adsorption promoting eluent while THF supresses adsorption in the system PS - silica gel. Smaller polymer coils have larger net column packing pore area accessible for adsorption and this may be responsible for above phenomenon.

Coupling exclusion with thermodynamic partition to increase selectivity of SEC separation. Isocratic elution with a single column packing

In this case, a two phase system must be generated, i.e. the composition must differ of eluent situated in the interstitial volume of column and in the quasi-stationary phase confined within the gel pores. This can be achieved, e.g. by using a binary liquid as eluent, and a column packing that preferentially absorbs one of the eluent components. Alternatively, homogeneously crosslinked or composite column packings (5) should be applied: In this latter case, the gel phase can be considered a „solution" of macromolecules in eluent thus creating a two-phase system necessary for partition of separated macromolecules (6).

The thermodynamic partition of solute molecules between the quasi-stationary gel phase and the mobile phase presents the additional separation mechanism. The systems are to be identified in which separated macromolecules prefer the gel phase, e.g. because of their increased solubility. As result, their retention volumes increase. Moreover, such chromatographic systems must be designed in which the solubility of analysed molecules in the eluent increases with their molar mass, e.g. due to end groups effects. In this way, the SEC separation selectivity can be enhanced. Similarly as in the case described in paragraph 1, the selectivity gain is large mainly with oligomers. For example, high separation selectivity was obtained for PEGs applying Sephadex LH-20 with benzene plus methanol mixed eluents (7). Sephadex LH-20 is a polar gel that preferentially absorbs methanol from eluent. This means that the quasi stationary eluent within gel contains more methanol than does the mobile phase. Methanol is a better solvent for PEG than benzene and, moreover, solubility of PEG in benzene falls with its decreasing molar mass. Therefore macromolecules prefer the methanol-rich gel phase over the benzene-rich mobile phase and their retention volumes rise. This effect increases with decreasing molar mass and the resulting calibration curve assumes a shape similar to curve **c** in Figure 1: The governing separation mechanism remains size exclusion but its selectivity has been enhanced by partition effects.

Several specific approaches to the characterization of complex polymers will be discussed in the following few paragraphs: The separation is suppressed of a particular kind of polymer chains according to a particular parameter (e.g. molar mass) to allow undisturbed separation of another kind of chains according to the same parameter:

Coupling exclusion with adsorption to eliminate separation according to molar mass

Several LC procedures combine exclusion with adsorption in a way that the separation according to the polymer molar mass is suppressed. These procedures are called liquid chromatography of macromolecules at the point of exclusion - adsorption transition (LC PEAT) (8). Three approaches to the LC PEAT were so far proposed:

Isocratic elution with a single packing: Liquid chromatography at the critical adsorption point, LC CAP. In the LC CAP, eluent is a liquid that slightly promotes polymer adsorption and sample is dissolved in eluent. At the critical adsorption point (CAP) the exclusion and adsorption effects mutually compensate so that no separation of polymer according to molar mass occurs (9-11) (Figure 1, line **d**). For the sake of clarity, LC CAP principle is illustrated in Figure 2. This situation can be employed for a very effective discrimination of oligomers according to their functionality that is according to the nature and/or number of functional groups per macromolecule (12). Similarly, chemically different constituents of binary polymer blends can be separated (13) with the help of LC CAP, including discrimination of parent homopolymers from the corresponding copolymers (14). The LC CAP method is even sensitive enough to separate macromolecules on the basis of their tacticity (15). Alternatively, the LC CAP principle can be applied to characterization of block- (13) and graft- (16) copolymers. In all these applications, one kind of polymer chains is eluted at the critical adsorption point. Simultaneously, another kind of polymer chains is eluted either according to liquid adsorption chromatographic mechanism (mainly in the case of oligomers) or according to size exclusion chromatographic mechanism (high polymers). The latter, „conventionally" eluted polymer chains are characterized by the procedures commonly used in liquid adsorption chromatography or size exclusion chromatography without interference from the „chromatographically invisible" chains eluted at the critical adsorption point, where effect of M on V_R vanishes.

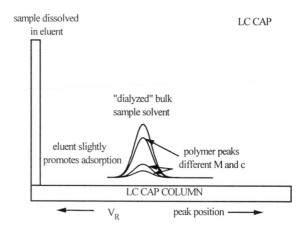

Figure 2. Schematic representation of LC CAP approach.

The LC CAP method possesses a large potential, mainly in the characterization of oligomers (12), as demonstrated also by MALDI-TOF MS (13,17). In the case of high polymers, the method, however, suffers from serious drawbacks (18,19). Therefore, the application limits of the LC CAP method must be tested for each system and the experimental conditions have to be carefully optimized. Many LC CAP problems are connected with enormous sensitivity of the critical adsorption point position to minute changes in both - eluent composition (including trace amounts of moisture) and temperature, as well as to variations in the physical structure of macromolecules. Further parameters affecting the critical adsorption point include column packing surface properties and/or column packing pore size (4). As result, the reproducibility of measurements is rather limited and it was found to be very difficult to apply the LC CAP method to polymers with broad molar mass distribution containing species with molar mass over 10^5 dalton. Other experimental problems connected with the LC CAP method include unexpectedly large peak broadening and even splitting (4,18), as well as difficulties with the sample detection: Most of LC CAP eluents are mixed systems containing at least two components. Macromolecules dissolved in mixed solvents usually exhibit preferential solvation (20). As result, initial sample solvent, bulk solvent and solvent in the vicinity of polymer coils, all differ in their composition. Preferential solvation complicates detection of macromolecules (21) and gives rise to the system peaks on chromatograms monitored by non-specific detectors (22). System peaks interfere with the peaks of macromolecules eluted at the critical adsorption point (23). Moreover, there may be found differences in the retention of macromolecules eluted together with their bulk solvent (e.g. homopolymers) and retention of polymer chains that were separated from their bulk solvent due to size exclusion effects of their „partner" chains in the block- or graft - copolymers) (24).

The problems generated by the bulk solvent zone can be suppressed by inserting a non-interactive column between the injection valve and the LC CAP column. This auxiliary column must have very narrow pores so that all macromolecules are excluded but bulk solvent permeates - and it is separated from polymer coils. In this way, a compensation between exclusion and adsorption can be reached that is not influenced with the presence of bulk solvent zone. Consequently, we arrive at the

Isocratic elution with a single packing: Liquid Chromatography at the theta exclusion-adsorption (LC TEA). In LC TEA (Figure 3), the dynamic equilibrium between preferentially solvated macromolecules and their bulk solvent is perturbed since polymer coils become surrounded by their initial solvent (eluent). Further (preferential) solvation can take place. Still, these reequilibration processes were not observed and the system peaks exhibited symmetrical shape (25). On the other hand, separation of macromolecules from their bulk solvent seems to diminish the sensitivity of retention of polymers near their theta exclusion - adsorption point (24). The effects of displacements and exchanges are reduced between sample bulk solvent and column packings surface layer of the preferentially adsorbed eluent component. As result, the exclusion-adsorption compensation can be better controlled. The strong interference between system peak and sample peak is suppressed.

Quasi-isocratic elution with a local microgradient using a single packing: Liquid chromatography of macromolecules under limiting conditions of adsorption (LC LCA). LC LCA is an emerging technique (7,26-28) based on a dynamic combination of exclusion and adsorption mechanisms. As with the LC CAP and LC TEA, macromolecules eluting under limiting conditions of adsorption exhibit a single retention volume independent of their sizes. The eluent is a liquid that slightly promotes adsorption of macromolecules while the sample solvent

184

strongly suppresses adsorption of macromolecules (desorption promoting liquid - a "DESORLI") (Figure 4). This is an important difference between LC LCA and LC CAP since the eluent is used as sample solvent in the latter case. Injected polymer has a tendency to elute faster than its initial solvent due to exclusion. Macromolecules, however, cannot leave the zone of their initial solvent otherwise they are retained within column by adsorption. When reached by the DESORLI zone, polymer will be desorbed and eluted again. Finally, an adsorption-desorption equilibrium is attained and macromolecules move near to the front of the DESORLI zone.

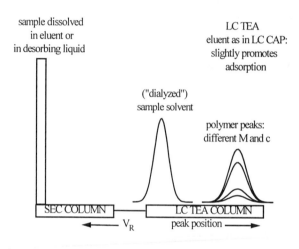

Figure 3. Schematic representation of LC TEA approach.

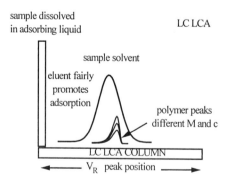

Figure 4. Schematic representation of LC LCA approach.

In comparison with LC CAP, the polymer peak broadening is usually less pronounced, the sensitivity of retention toward eluent composition changes is reduced and the limit of applicable molar masses is increased in the case of LC LCA (8,27). On the other hand, an extensive peak splitting was observed in some LC LCA systems (8,27) and the applicability of the column packing - adsorbing eluent - desorbing sample solvent must be carefully tested.

LC LCA can be employed for the separation of polymer mixtures with constituents differing in their affinity to the column packing due to their different chemical nature or physical structure (8,15,26,27). The applications of LC LCA include also water soluble polymers (28) and possibly also copolymers (29,30). In the latter case, however, macromolecules of particular compositions leave the zone of their initial DESORLI solvent and may be separated according to both their molar mass and chemical composition. Consequently, the procedure is de facto isocratic liquid adsorption chromatography in which one copolymer constituent is only weakly adsorbed within column packing. LC LCA approach probably only enables to identify the appropriate experimental conditions, i.e. eluent composition, temperature - or column packing for such isocratic LAC elution.

Coupling exclusion with solubility to eliminate separation according to polymer molar mass. Quasi-isocratic elution with local microgradients using a single packing: Liquid chromatography of macromolecules under limiting conditions of solubility (LC LCS)

This method represents a dynamic combination of exclusion and precipitation / dissolution mechanisms resulting in retention volumes of macromolecules independent of their molar masses (28,31-33). Eluent is a **weak** nonsolvent for macromolecules injected in a thermodynamically **good** solvent. In this way, the sample solvent forms a „microgradient zone" that moves along the column. Similar as in the case of LC LCA, polymer tends to elute more rapidly than its initial solvent due to exclusion. However, macromolecules cannot leave the zone of their initial solvent otherwise they would precipitate and stop moving. Precipitated polymer will redissolve when reached with the zone of its solvent and starts to elute again. If precipitation and redissolution are properly balanced, macromolecules of all sizes travel at about the same velocity and the calibration curve like **d** in Figure 1 is generated.

Solubility of macromolecules strongly depends on their molar masses. It may happen that polymer that exhibits a vertical calibration curve due to its insolubility in the eluent (LCS) becomes soluble at its lower molar masses. Still, the molar mass independent retention may be maintained because of the LCA principle (28). We speak about a hybrid elution mechanism.

It is obvious that adsorption of polymer species on the column packing surface will often affect the position of retention volumes at the limiting conditions of solubility.

LC LCS/LCA can be applied to the separation of polymer blend components differing in their solubility due to diversity in their chemical nature or physical structure. Besides mixtures of chemically different homopolymers (31,32) also blends of poly(methyl methacrylate)s of different tacticities were discriminated with the LC LCS/LC LCA method (33): One component of a polymer mixture has been eluted under LCS/LCA conditions while another, better soluble polymer has been eluted in the SEC mode and it could be characterized in the conventional way.

Coupling SEC with full adsorption/desorption processes to successively separate chemically different species according to their molar mass (FAD/SEC): Column packing eluent and/or switching procedure

This new technique (34-38) resembles on-line solid phase extraction. The constituents of a polymer blend under investigation are completely retained within an especially designed full adsorption-desorption column (FAD). In the following steps the trapped constituents are successively displaced, e.g. by stepwise eluent changes, and are transported into an on-line SEC system for further characterization. Two-component (34, 35) or multicomponent (36, 37) polymer blends can be separated and independently characterized by the above FAD/SEC combination. The mixtures of very similar macromolecules such as various esters of polymethacrylic acid can be easily discriminated in this way, too (36). The method possesses a remarkable potential also for characterization of various kinds of copolymers that are built of monomers with different adsorptive properties (36), as well as for separation of oligomers (38).

Coupling LC CAP with temperature gradient to enhance selectivity of separation

This new separation technique was proposed by Chang et al (39) under the name temperature gradient interactive chromatography (TGIC). It is based on the high sensitivity of polymer retention toward temperature variations in the area of the critical adsorption point (18,19). When applying a stepwise or continuous temperature gradient to the system which is in the vicinity of its critical adsorption point, the changes strongly increase of retention volumes with polymer molar masses, that is the selectivity of separation dramatically rises. The procedure can be directly combined with conventional SEC for effective discrimination of polymer mixtures (40). The non interacting macromolecules (one constituent of polymer blend) are separated with SEC mechanism while the interacting macromolecules (another constituent) elute under TGIC conditions: their retention volumes increase with increasing molar mass. In the appropriately designed eluent/column packing system, polymer which does not interact with column packing is separated according to the SEC mechanism. At the same time the interacting macromolecules are separated according to their molar mass with help of the temperature controlled adsorption mechanism.

Multidimensional liquid chromatography

In this case, SEC, LAC, LC CAP, supercritical fluid chromatography, LC TEA, LC LCA or LC LCS are on-line combined to effectively separate complex polymer systems. It is evident that SEC is a very important component in these „higher degree of coupling" procedures. The value of information obtained by the above couplings can be further enhanced applying detectors that continuously monitor molar mass and/or structure of macromolecules, e.g. mass spectrometers (13,17) or NMR detectors (41).

Another candidate for coupling with SEC represents full precipitation - continuous gradient redissolution liquid chromatography called also high performance precipitation liquid chromatography or gradient polymer elution chromatography. In its high performance version, this method was proposed Glöckner (42) and further developed e.g. by Klumperman et al. (43). The eluent is a nonsolvent for macromolecules that are injected into the column in their good solvent. The conditions are chosen so that polymer fully precipitates near the column entrance. Subsequently, eluent with gradually increasing dissolution power is introduced into column and macromolecules successively redissolve and travel along the LC column. Polymer solubility depends on its chemical composition, molar mass and physical structure. This means that even if the column packing is chosen in which no exclusion processes are to be expected, the resulting separation is often a complicated function of at least two of above-mentioned parameters. The attempts are made to suppress the molar mass dependence of elution by an appropriate choice of eluent (43) or column packing.

Further opportunity for couplings offers full adsorption-continuous gradient desorption liquid chromatography (44-46). The eluent strongly promotes adsorption of macromolecules on the given column packing and the injected polymer is fully retained near the column inlet by adsorption. Next, macromolecules of different compositions (e.g. fractions of a random copolymer) are successively desorbed by eluent with continuously increasing content of an appropriate desorbing liquid. In effect, macromolecules are eluted mainly in dependence on their composition. Microporous column packing is applied to suppress the size exclusion effects. The fractions from adsorption column can be further separated by SEC applying a stop-and-go approach in either adsorption or exclusion column.

The full adsorption-continuous gradient desorption method resembles FAD procedure. However, in the latter case a stepwise displacement (eluent switching) is employed rather than a continuous gradient and the actual chromatographic separation is not expected to take place within FAD column.

As demonstrated, several coupled LC techniques were proposed but so far none of them can be considered established and widely used. New LC combinations keep emerging since recently and the development in the coupled LC methodology for separation of complex polymers is anticipated to be very fast in the near future.

The general problem of most coupled LC separation methods include the quantitative data processing. In contrast with SEC, each particular system must be carefully calibrated because no „universal calibration" is valid and the „absolute" detectors like light scattering or viscosity detectors are difficult to apply. This is especially the case for oligomers if their components cannot be base line separated and for high polymers exhibiting non-uniform composition and/or eluted with the

188

continuous solvent gradient. Evidently the development of new separation methods must be accomplished with novel detection and data processing procedures.

Acknowledgement

This work was supported by the Slovak Grant Agency VEGA (Project No. 2/4012/97) and partially also by the US-SK Scientific Technical Cooperation Grant 007-95.

Literature Cited

1. Berek, D.; Bakoš, D. *J. Chromatogr.* **1974,** *91,* 237
2. Novák, L.; Matejeková, V.; Berek, D. Die niedrig Oligomere in den nichtionogenen Tensiden auf Basis der Nonylphenol - Polyglykoläther, IN: Chemie, physikalische Chemie und Anwendungstechnik der grenzflächenaktiven Stoffe, C. Hansen Verlag, Munich **1973,** *1,* 453
3. Chiantore, O. *J. Liq. Chromatogr.* **1984,** *7,* 1
4. Berek, D.; Janèo, M. *J. Polym. Sci. A: Polym. Chem.* **1998,** *36,* 1363
5. Petro. M.; Berek, D. *Chromatographia* **1993,** *37,* 549, and Petro, M.; Berek, D.; Novák, I. React. Polym. **1994,** *23,* 173
6. Audebert, R. *Polymer* **1979,** *20,* 1561
7. Berek, D. *Chromatography of polymers and polymers in chromatography,* DSc Thesis, Bratislava 1991
8. Berek, D. *Liquid Chromatography of Macromolecules at the Point of Exclusion-Adsorption Transition, Macromol. Symp.,* in press
9. Belenkii, B.G.; Gankina, E.S.; Tennikov, M.B.; Vilenchik, L.Z. *Dokl. Akad. Nauk* SSSR (Moscow) **1976,** *231,* 1147 and *J. Chromatogr.* **1978,** *147,* 99
10. Skvortsov, A.M.; Gorbunov, A.A. *Vysokomol. Sojed. (Moscow)* **1979,** *A21,* 339
11. Gorbunov, A.A.; Skvortsov, A.M. *Adv. Colloid Interface Sci.* **1995,** *62,* 31
12. Entelis, S.G.; Evreinov, V.V. ; Gorshkov, A.V. *Adv. Polym. Sci.* **1987,** *76,* 129
13. Pasch, H. *Macromol. Symp.* **1996,** *110,* 107
14. Jančo, M.; Berek, D.; Onen, A.; Fischer, C.-H.; Yagci, Y.; Schnabel, W. *Polym. Bull.* **1977,** *38,* 681
15. Berek, D.; Jančo, M.; Hatada, K.; Kitayama, T.; Fujimoto, N. *Polym. J.* **1997,** *29,* 1029
16. 16.Murgašová, R.; Berek, D.; Capek, I. *ICPG Newsletter* **1995,** *26,* 12 and Capek, I.; Murgašová, R.; Berek, D. *Polym. Int.* **1997,** *44,* 174
17. Trathnigg, B.; Maier, B.; Schulz, S.; Krüger, R.-P.; Just, U. *Macromol. Symp.* **1996,** *110,* 231
18. Philipsen, H.J.A.; Klumperman, B.; van Herk, A.M.; German, A.L. *J. Chromatogr. A* **1996,** *A727,* 13
19. Berek, D. *Macromol. Symp.* **1996,** *110,* 33
20. Casassa, E.F.; Eisenberg, H. *Adv. Protein Chem.* **1964,** *19,* 287
21. Trathnigg, B.; Thamer, D.; Yan, X.; Maier, B.; Holzbauer, H.-R.; Much, H. *J. Chromatogr.* **1994,** *A605,* 47

22. Berek, D.; Bleha, T.; Pevná, Z. *J. Chromatogr. Sci.* **1976,** *14,* 560
23. Pasch, H.; Rode, K. *Macromol. Chem. Phys.* **1996,** *197,* 2691
24. Berek, D. Unpublished results
25. Berek, D.; Bleha, T.; Pevná, Z. *J. Polym. Sci., Polym. Lett. Ed.* **1976,** *14,* 323
26. Berek, D. *Coupled Procedures in Liquid Chromatography of Macromolecules, 5th Latin American Polymer Symposium - SLAP '96,* Mar del Plata 1996, p. 37
27. Berek, D.; Hunkeler, D. *Liquid Chromatography of Macromolecules under Limiting Conditions of Adsorption, J. Polym. Sci. A: Polym.* Chem. submitted
28. Bartkowiak, A.; Hunkeler, D.; Berek, D.; Spychaj, T., *J. Appl. Polym. Sci.,* in press
29. Bartkowiak, A.; Hunkeler, D. *Polym. Mater. Sci. Eng.* **1998,** *78,* 59
30. Bartkowiak, A.; Hunkeler, D. *Liquid Chromatography under Limiting Conditions: A Tool for Copolymer Characterization,* this book, T. Provder, ed., ACS Press, Washington, D.C., 1998
31. Hunkeler, D.; Macko, T.; Berek, D. *ACS Symp. Ser.* **1993,** *521,* 90
32. Hunkeler, D.; Jančo, M.; Guryanova, V.V.; Berek, D. *ACS Adv. Chem. Ser.* **1995,** *247,* 13
33. Berek, D.; Jančo, M.; Kitayama, T.; Hatada, K. *Polym. Bull.* **1994,** *32,* 629
34. Jančo, M.; Berek, D.; Prudskova, T. *Polymer* **1995,** *36,* 3295
35. Jančo, M.; Prudskova, T.; Berek, D. *Int. J. Polym. Anal. Charact.* **1997,** *3,* 319
36. Nguyen, S.H.; Berek, D, *„Full Adsorption-Desorption/SEC Coupling in Characterization of Polymers",* this book, T. Provder, ed., ACS Press, Washington, D.C., 1998
37. Nguyen, S.H.; Berek, D. *Chromatographia,* in press.
38. Trathnigg, B.; Nguyen, S.H.; Berek, D. in preparation
39. Lee, H.C.; Ree, M.; Chang, T. *Polymer* **1996,** *37,* 5747
40. Lee, H.C.; Chang, T. *Macromolecules* **1996,** *29,* 7294
41. Hatada, K.; Ute, K.; Nishimura, T.; Kashiyama, M.; Fujimoto, N. *Polym. Bull.* **1990,** *23,* 549
42. Glöckner, G. *Gradient HPLC of Copolymers and Chromatographic Crossfractionation,* Springer, Berlin (1991)
43. Klumperman, B.; Cools, P.; Philipsen, H.; Staal, W. *Macromol. Symp.* **1996,** *110,* 1
44. Mori, S. *Macromol. Symp.* **1996,** *110,* 87
45. Teramachi, S. *Macromol. Symp.* **1996,** *110,* 217
46. Sato, H.; Ogino, K.; Darwint, T.; Kiyokawa, I. *Macromol. Symp.* **1996,** *110,* 177

Chapter 13

Two-Dimensional Liquid Chromatography of Functional Polyethers

Bernd Trathnigg, Manfred Kollroser, M. Parth, and S. Röblreiter

Institute of Organic Chemistry, Karl-Franzens-University at Graz, A-8010 Graz, Heinrichstrasse 28, Austria

Two-dimensional liquid chromatography allows full characerization of complex polyethers, such as fatty alcohol ethoxylates (FAE), macromonomers and EO-PO-block copolymers. Typically, liquid chromatography under critical conditions is used as the first dimension. In the second dimension, size exclusion chromatography (SEC) or liquid adsorption chromatography (LAC), typically in gradient mode, can be applied.

Poly(ethylene glycol)s with functional end groups and block copolymers of ethylene oxide (EO) and propylene oxide (PO) are used in many fields. According to the hydrophilic nature of the polyoxyethylene chain, they may be used as nonionic surfactants, the most important of which are fatty alcohol ethoxylates (FAE). As these products consist of different polymer homologous series, their full characterization requires a two-dimensional separation (according to functionality and molar mass distribution). A similar situation is found in EO-PO block copolymers, in which each molecule may contain different numbers of EO and PO units.

Basically, different modes of liquid chromatography, which can be applied in the analysis of polymers, may also be combined to achieve multidimensional separations:

- Size exclusion chromatography (SEC) separates according to molecular size (not actually molar mass !)[1, 2]. It is always performed in isocratic mode, typically in pure solvents.

- Liquid chromatography at the critical point of adsorption (often also called LC under critical conditions; LCCC)[3-8] is run at a special temperature and mobile phase composition, at which all chains with the same repeating unit elute at the same elution volume (regardless their length), which means, that the polymer chain (or one block) becomes chromatographically invisible. In this case, a separation according to a structural units other than the repeating unit (end groups, one block in a copolymer etc.) can be achieved. LCCC is also run under isocratic conditions, but typically in mixed mobile phases.

- Liquid adsorption chromatography (LAC) separates according to chemical composition and to molar mass[9]. In principle, LAC can be performed using isocratic or gradient elution, but samples with higher molar mass typically require gradients[10-12].

An excellent review on the separation of poly(ethylene glycol)s and their derivatives by different techniques is given by Rissler[13].

In the case of FAE and EO-PO- block copolymers, the most feasible approach is a combination of LCCC on a reversed-phase column (under critical conditions for polyoxyethylene) as the first dimension and a separation of the fractions thus obtained according to the number of EO units.

In previous papers[1, 14] we have shown, that FAE can be separated according to the alkyl end groups on an ODS packing in methanol-water; the pure homologous series thus obtained were analyzed by SEC in chloroform[15].

Using a combination of density and RI detection, the chemical composition along the MMD can be determined, as is shown in another chapter of this book.

While a quantitatively accurate analysis of the pure polymer homologous series in the second dimension is rather easy, the main problem in the analysis of such complex samples is quantitation of the first dimension, where unknown amounts of a fraction with unknown chemical composition elute in one peak together with an unknown amount of one component of the mobile phase (due to preferential solvation of the polymer chains). With coupled density and RI detection in both dimensions, a quantitatively accurate mapping of FAE was achieved, as is shown schematically in Figure 1.

A full separation of all oligomers was, however, not achieved because of the limited separation power of SEC.

This appeared to be possible by replacing SEC by LAC: as has been shown by several authors [16-20], ethoxylates can be separated by normal phase LAC. Hence we have tried to combine LCCC as the first and normal phase LAC as the second dimension in order to achieve a mapping of FAE with a full resolution of all oligomers.

Experimental:

These investigations were performed using the density detection system DDS70 (CHROMTECH, Graz, Austria), which has been developed in our group. Each system was connected to a MS-DOS computer via the serial port.

192

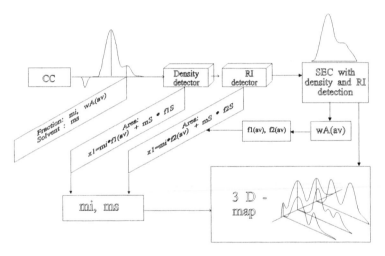

Figure 1:   Principle of 2D-LC with multiple detection

Data acquisition and processing was performed using the software package CHROMA, which has been developed for the DDS 70. The columns and density cells were placed in a thermostatted box, in which at temperature of 25.0°C was maintained for all measurements.

LCCC was performed in methanol-water 90:10 (w/w) (both solvents from Merck, HPLC grade) on a semi-preparative column (10x250 mm) packed with Spherisorb ODS2 (5 µm, 80 Å) from PhaseSep (Deeside, Clwyd, UK) at a flow rate of 2.0 ml/min, which was maintained with a ISCO 2350 HPLC pump (from ISCO, Lincoln, Ne, USA).
An Advantec 2120 fraction collector was used in the fractionations (from Advantec, Dublin, CA, USA). A SICON LCD 201 RI detector was combined with the DDS70.

SEC measurements were performed in chloroform (HPLC grade, Rathburn) at a constant flow rate of 1.0 ml/min, which was maintained by a Gynkotek 300C HPLC pump. Samples were injected using a VICI injection valve (Valco Europe, Switzerland) equipped with a 100 µl loop, the concentration range was 4 - 8 g/l. A column set of 4 Phenogel columns (2*500 Å+2*100 Å, 5 µm, 7.8 x 300 mm each), was used for all measurments. In SEC, an ERC 7512 RI detector (ERMA,) was combined with the DDS70.
The SEC calibrations were obtained using pure oligomers of EO (Fluka, Buchs, Switzerland)).
Chemical composition along the MMD was determined from density and RI detection using the software CHROMA.

In **LAC** measurements, the mobile phase was delivered by two JASCO 880 PU pumps (from Japan Spectrosopic Company, Tokyo, Japan), which were coupled in order to provide gradients by high pressure mixing. The solvents (acetone and water) were HPLC grade (from Promochem, Wesel, Germany).
In gradient elution, mobile phase A was pure acetone, mobile phase B was acetone-water 80:20 (w/w). The following gradient profile was used: start 100 % A, then in 50 min to 100 % B, 4 min constant at 100 % B, then within 1 min back to 100 % A.
A SEDEX 45 ELSD (Sedere, France) was connected to the DDS 70. Nitrogen was used as carrier gas, and the pressure at the nebulizer was set to 2.0 bar, the temperature of the evaporator to 30°C . The following columns were used, which were connected to two column selection valves (Rheodyne 7060, from Rheodyne, Cotati, CA, USA) :
a)   Spherisorb S3W, 3µm, 80 Å, 100 x 4.6 mm
b)   Spherisorb S5W, 5 µm, 80 Å, 250 x 4.6 mm
Samples were injected manually using a Rheodyne 7125 injection valve (from Rheodyne, Cotati, CA, USA) equipped with an 50 µl loop.

Two series of FAE samples were used in these investigations:
Brij 30 and Brij 35 were purchased from Fluka (Buchs, Switzerland), Dehydol LT8 were provided by Henkel (Germany).

The specifications given by the producers were as follows:
Brij 30: Polyethylene glycol dodecyl ether, main component: tetraethylene glycol dodecyl ether; Brij 35: Polyethylene glycol dodecyl ether, main component: trikosaethylene glycol dodecyl ether; Dehydol LT8: fatty alcohol C12-C18 + 8 EO.

Results and discussion:
In Figure 2, a separation of Dehydol LT8 by LCCC is shown: the individual homologous series are clearly separated.
When the fractions thus obtained are analyzed by SEC with coupled density and RI detection, the MMD and the chemical composition is obtained, as is shown in Figure 3.
Such a combination of LCCC and SEC with coupled density and RI detection in both dimensions has been successfully applied in the characterization of fatty alcohol ethoxylates[2]. More details on this procedure are given in another chapter of this book.
A full separation of the individual oligomers is, however, not achieved, because the separation power of SEC is not sufficient.
On the other hand, the interaction of higher oligomers with the stationary phase is typically too strong in liquid adsorption chromatography, as can be seen from Fig.4, which shows a chromatogram Dehydol LT8, which was obtained on a normal phase column with isocratic elution.
While under these conditions the lower oligomers are not really separated, the higher oligomers appear as very broad peaks, which can hardly be integrated, and some more are obviously not even eluted.
Using gradient elution, a full separation of the individual oligomers is achieved, as can be seen from Figure 5: when fractions from the separation shown in Fig.2 were analyzed by gradient LAC, all oligomers could be completely separated. Of course, the combination of density and RI detector had to be replaced by an Evaporative Light Scattering Detector (ELSD)[21-23].
In Figure 6 and 7, a two-dimensional separation of Brij 30 is shown: fractions from LCCC (Fig.6) were separated by gradient LAC with ELSD.
In Figure 6, fractions Nr. 3, 6, and 9 contain the fatty alcohols C12OH, C14OH, and C16OH, fractions 4,7, and 11 contain the corresponding ethoxylates. PEG would appear in the solvent peak (fraction 1), but was not found in this sample.
In this case, the fatty alcohols elute in LCCC as sufficiently resolved peaks in front of the ethoxylates, Hence one can determine them quantitatively already in LCCC, and analyse only the ethoxylates by gradient LAC: The first peaks in Fig. 7 are thus the monoethoxylates.
Under the same conditions, we could separate even higher oligomers of FAE. As can be seen from Figures 8 and 9, a full separation of oligomers is achieved for such a fraction from LCCC.
As has already been shown in a previous paper, this sample also contains small amounts of poly(ethylene glycol), which is eluted in LCCC in the solvent peak.

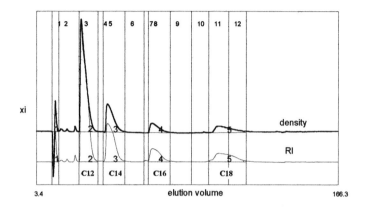

Figure 2:   Chromatogram of   Dehydol LT8, as obtained on
Spherisorb ODS2 (5 μm, 300 x 10 mm) in methanol-water 90:10
(w/w) at a flow rate of 2.0 ml/min. Detection: density and RI

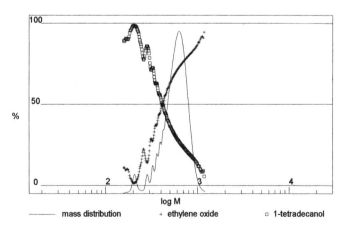

Figure 3:   Molar mass distribution (MMD) and chemical composition of the C14 fraction of Dehydol LT8 (from Figure 2), as
obtained by SEC with coupled density and RI detection.

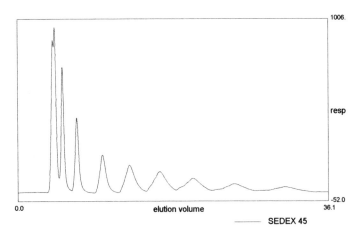

Figure 4: Chromatogram of Dehydol LT 8, as obtained with isocratic elution on Spherisorb S5W (5 μm, 80 Å, 250 x 4.6 mm) in acetone-water 99:1 (w/w) at a flow rate of 1.0 ml min. Detection: RI

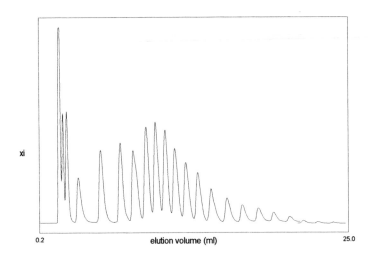

Figure 5: Gradient elution of the C12 fraction (see Fig 2) of Dehydol LT8, as obtained on Spherisorb S3W

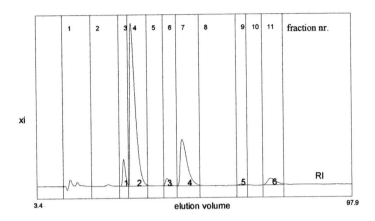

Figure 6: LCCC of Brij 30 with fraction limits (chromatographic conditions as in Figure 2).

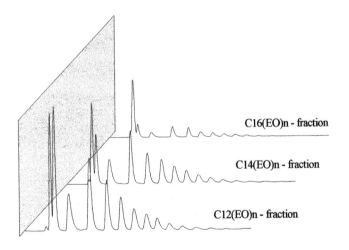

Figure 7: Mapping of Brij 30 by coupled LCCC (Fig. 6) and gradient elution on Spherisorb S3W with ELSD.

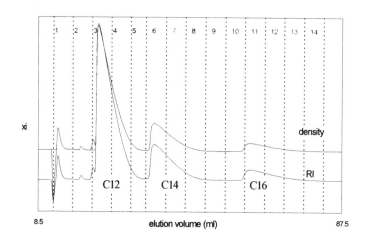

Figure 8:   LCCC of Brij 35 (chromatographic conditions as in Figures 2 and 6)

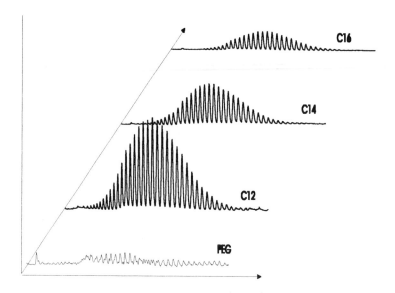

Figure 9: Chromatograms of the individual   fractions of Brij 35 (see Figure 7), as obtained by gradient elution on Spherisorb S3W with ELS detection. Chromatographic conditions as in Figure 4.

As can be seen from Figures 7 and 9, a full separation of all oligomers can be achieved by the combination of LCCC with gradient LAC on a normal phase column.

There is, however, concern about the quantitation of results: the response of the ELSD is typically non-linear, the response factors of individual oligomers are different for the ELSD, and they depend not only on the operating conditions[17, 24-27], but also on the composition of the mobile phase, in which they are eluted!

As is shown in another chapter of this book, this problem can be overcome by combining the ELSD with the density detector, from which the actual mobile phase composition for each peak is easily obtained.

Conclusions:

A combination of LCCC with coupled density and RI detection in the first dimension and gradient HPLC on a normal phase with ELSD in the second dimension yields a full resolution of oligomers in FAE.

There are, however, still many questions concerning quantitation. First of all, the linear range of the ELSD is very narrow. Its response can, however, be expressed by an exponential function.

Moreover, the dependence of its response on molar mass and chemical composition of the sample is not always clear.

This means, that an ELSD must be calibrated very carefully in order to yield reliable results, and the operating conditions have to be optimized and controlled for each measurement.

Acknowledgement

Financial support by the Austrian Academy of Sciences (Grant OWP-53) is gratefully acknowledged.

References:

1.) Trathnigg, B., Thamer, D., Yan, X., Maier, B., Holzbauer, H. R. and Much, H.; *J Chromatogr a* **1993,** *657,* 365-375.

2.) Ysambertt, F., Cabrera, W., Marquez, N. and Salager, J. L.; *J Liq Chromatogr* **1995,** *18,* 1157-1171.

3.) Pasch, H. and Zammert, I.; *J Liq Chromatogr* **1994,** *17,* 3091-3108.

4.) Kruger, R. P., Much, H. and Schulz, G.; *J Liq Chromatogr* **1994,** *17,* 3069-3090.

5.) Hunkeler, D., Macko, T. and Berek, D.; *ACS Symp Ser* **1993,** *521,* 90-102.

6.) Gorshkov, A. V., Much, H., Becker, H., Pasch, H., Evreinov, V. V. and Entelis, S. G.; *J Chromatogr* **1990,** *523,* 91-102.

7.) Skvortsov, A. M. and Gorbunov, A. A.; *Journal Of Chromatography* **1990,** *507,* 487-496.

8.) Skvortsov, A. M., Gorbunov, A. A., Berek, D. and Trathnigg, B.; *Polymer* **1998,** *39,* 423-429.

9.) Trathnigg, B., Thamer, D., Yan, X. and Kinugasa, S.; *J Liq Chromatogr* **1993,** *16,* 2439-2452.

10.) Rissler, K., Kunzi, H. P. and Grether, H. J.; *J Chromatogr* **1993,** *635,* 89-101.
11.) Rissler, K., Fuchslueger, U. and Grether, H. J.; *J Liq Chromatogr* **1994,** *17,* 3109-3132.
12.) Marquez, N., Anton, R. E., Usubillaga, A. and Salager, J. L.; *J Liq Chromatogr* **1994,** *17,* 1147-1169.
13.) Rissler, K.; *J Chromatogr A* **1996,** *742,* 1-54.
14.) Trathnigg, B., Thamer, D., Yan, X., Maier, B., Holzbauer, H. R. and Much, H.; *J Chromatogr A* **1994,** *665,* 47-53.
15.) Trathnigg, B. and Kollroser, M.; *Int.J.Polym.Anal.Char.* **1995,** *1,* 301-313.
16.) Desbene, P. L., Portet, F. I. and Goussot, G. J.; *J Chromatogr A* **1996,** *730,* 209-218.
17.) Miszkiewicz, W. and Szymanowski, J.; *J Liq Chromatogr Relat Techno* **1996,** *19,* 1013-1032.
18.) Jandera, P., Urbanek, J., Prokes, B. and Blazkovabrunova, H.; *J Chromatogr A* **1996,** *736,* 131-140.
19.) Ibrahim, N. M. A. and Wheals, B. B.; *J Chromatogr A* **1996,** *731,* 171-177.
20.) Trathnigg, B., Maier, B., Gorbunov, A. and Skvortsov, A.; *J Chromatogr A* **1997,** *791,* 21-35.
21.) Mengerink, Y., Deman, H. C. J. and Van der Wal, S.; *J Chromatogr* **1991,** *552,* 593-604.
22.) Dreux, M. and Lafosse, M.; *Analusis* **1992,** *20,* 587-595.
23.) Lafosse, M., Elfakir, C., Morin-Allory, L. and Dreux, M.; *HRC High Res Chromatogr* **1992,** *15,* 312-318.
24.) Van der Meeren, P., Van der Deelen, J. and Baert, L.; *Anal Chem* **1992,** *64,* 1056-1062.
25.) Caron, I., Elfakir, C. and Dreux, M.; *J Liq Chromatogr Relat Techno* **1997,** *20,* 1015-1035.
26.) Koropchak, J. A., Heenan, C. L. and Allen, L. B.; *J Chromatogr A* **1996,** *736,* 11-19.
27.) Trathnigg, B. and Kollroser, M.; *J Chromatogr A* **1997,** *768,* 223-238.

Chapter 14

Liquid Chromatography under Limiting Conditions: A Tool for Copolymer Characterization

A. Bartkowiak and D. Hunkeler

Laboratory of Polymers and Biomaterials, Swiss Federal Institute of Technology, CH-1015 Lausanne, Switzerland

Liquid chromatography under limiting conditions of solubility (LC LCS), a method in which the enthalpic and entropic separation mechanism are perfectly compensated over a large molecular weight range, has been used for the characterization of random copolymers. By eluting macromolecules such as polystyrene-co-methylmethacrylate at the LC LCS for one of the homopolymers (58 vol% THF/42 vol% n-hexane for polymethylmethacrylate) one can distinguish the copolymers according to the molar fraction of polystyrene groups. This can be accomplished through calibrating either according to retention volume shifts or peak area. Such LC LCS conditions can subsequently be coupled with a classical SEC separation in order to deconvolute copolymer composition and molecular size distributions.

Over the past two decades there have been a series of chromatographic methods which combine enthalpic and entropic separations [1-4]. Belenkii was the first, in a thin layer chromatography arrangement, to utilize a binary eluent combination to equate the free energy of adsorption with that of exclusion. Given this, "critical conditions" were identified where macromolecules eluted with a retention volume independent of molar mass [1]. Other chapters in this book [5] describe the methods which are now under development including liquid chromatography at the point of the elution-adsorption isotherm (LC-PEAT), as well as liquid chromatography under limiting conditions of adsorption (LC LCA) and solubility (LC LCS). These chapters are complimented by recent reviews [2,6,7]. It is advances in the latter two phenomena which will be described in this chapter, with a particular emphasis on the characterization of random copolymers with high molar masses.

In contrast to LC-PEAT, where the binary eluent involves two solvents for the polymer probe and the composition at the elution-adsorption transition is very sensitive to fractional changes in the solvent ratios, limiting conditions are achieved by combining a non-solvent for the polymer as a component of the mobile phase. Indeed, the thermodynamic quality of the solvent in LC LCS is such that the polymer cannot dissolve in the mobile phase. Given this, polymers are injected in a thermodynamically good solvent. The resulting mechanism involves a continuous process of elution, adsorption and redissolution in which the macromolecules move at a velocity faster then the injection zone, encounter the mobile phase adsorb onto the sorbent surface. As the injection zone reaches the adsorbed polymer, the chains are subsequently redissolved. The net result is the elution of the polymer, independent of its molar mass, just in front of the system peak [8]. It has been shown through cloud point curve measurements [9] that the retention independent composition can occur when the operative forces within the column are a combination of exclusion and adsorption (i.e. LC LCA) or exclusion, adsorption and solubility (LC LCS). These two cases are depicted in Figure 1. Figure 2 shows example of LC LCA identified for polystyrene in THF/n-hexane over a polystyrene/divinylbenzene sorbent. By comparison for the same mobile and stationary phase combination we have observed for the polymethylmethacrylate the LC LCS. Clearly in this case the calibration curve is in the insoluble portion of the cloud point curve (Figure 3). Recently the first hybrid LC LCA-LCS system have been identified [4] for a water soluble polymer (polyacrylamide), as is shown in Figure 4.

Although the critical condition, or LC-PEAT, methodologies have been theoretically modeled [10] and experimentally shown to work well for the characterization of functional groups [11], the molecular weight limit for the exclusion-adsorption point is approximately one-hundred thousand daltons. This should not be surprising since the free energies of exclusion and adsorption have quite different molecular weight dependencies. Recently LC-LCA has been shown to be effective in the characterization of polymer blends [6] as well as the tacticity of copolymers [12]. Interestingly, the use of a non-solvent as a component of the mobile phase enables the limiting condition of adsorption or solubility to exceed one million daltons [13]. In the case of LC LCS the use of a second enthalpic separation mechanism, based on solubility, seems to permit the compensation of molar mass dependencies of exclusion and adsorption/solubility over four orders of magnitude [7,8]. This lack of an upper provides hope that these methods will be suitable for the characterization of high copolymers. While experimental data to date has shown the ability to characterize one block of a copolymer, at a limiting condition for the second block, this chapter will report the first finding on the characterization of random copolymers with LC LCS. Clearly this is quite important, since the ability to eliminate a separation according to size permits one to fractionate polymers according to their composition distributions. If an SEC column is added in series the polymer can subsequently be separated according to molecular size. Therefore, the deconvolution of the composition and molecular size distributions is possible with LC LCS, under certain conditions, as will be discussed.

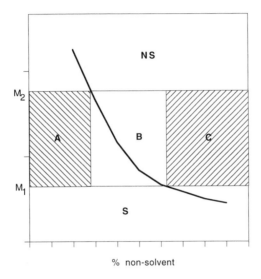

% non-solvent

Figure 1. A schematic plot of the solubility of polymer standards in a mixed eluent (solvent plus non-solvent) system, in interactive liquid chromatography experiments. In domain A, adsorption is the operative enthalpic mechanism which is balanced with exclusion (LC-LCA). In domain C the polymer solvent solubility dominates the enthalpy. Domain B is a hybrid where the entropic exclusion forces are balanced by the adsorption and solubility. Note that M_1 an M_2 represent the range where the retention volume is independent of the polymer molecular weight.

204

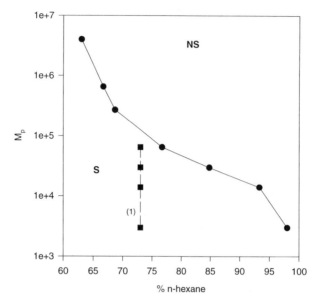

Figure 2. A cloud point curve for polystyrene (●) in THF/n-hexane. Measurements were performed at a polymer concentration of 1.0 mg/mL. Line 1 designates the LC LCA point (■), which is clearly in the soluble domain (S).

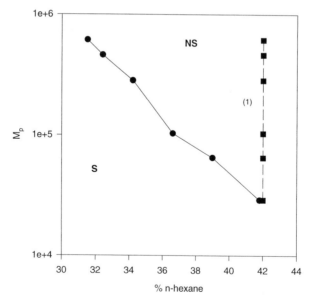

Figure 3. A cloud point curve for polymethylmethacrylate (●) in THF/n-hexane. Measurements were performed at a polymer concentration of 1.0 mg/mL. Line 1 (■) designates the LC LCS point, which is clearly in the insoluble domain (NS).

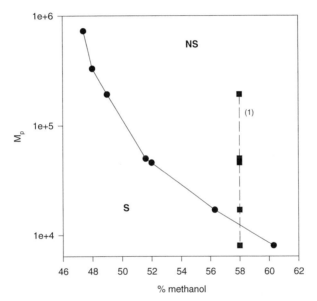

Figure 4. A plot of the solubility of polyacrylamide (PAM) standards (●) in a mixed eluent (0.05 M aqueous Na₂SO₄ plus methanol) system. PAM of various molecular weights are soluble to the left of cloud point curve (S zone), and insoluble to the right of solid line (NS zone). Line 1 (■) represents the retention independent MW condition (LC-LCS for molecular weights above 2 10⁴ daltons) for PAM observed on a polyhydroxymethacrylate gel (data from reference [8]).

Experimental

Liquid Chromatograph. An L-6000 (Hitachi Instruments, Tokyo, Japan) isocratic pump coupled with an Hitachi L-4000 UV detector operating at a wavelength of 234 nm were utilized in all experiments. A Rheodyne type 7725i valve (Coati, CA, USA) with an injection loop of 20 mL was employed. Chromatograms were collected on a Pentium computer running LaChrom D-7000 Multi HPLC Manager Software (Merck, Germany). The separations involved 1.5 mL/min flowrate, a solute concentration of 1.0 mg/mL with a 2 cm tubing (500 mm ID) connection length between the valve and column.

Mobile and Stationary Phases. Spectranalyzed grade THF (Fisher, Norcross, GA, USA and Merck, Switzerland) and HPLC grade n-hexane (Fisher and Merck, Switzerland) were used as received. A Shodex (JM Science, Grand Island, NY, USA) linear GPC 806L column (0.8 x 30 cm) packed with 10 μm polystyrene-co-divinylbenzene particles was employed for all experiments. Experiments were carried out at 25 ±0.1 °C in a Hitachi L-7300 column oven.

Polymers. Polystyrene standards with a molecular weight range of 370-1,400,000 were obtained from American Polymer Standards Corporation (Mentor, Ohio, USA). Atactic polymethylmethacrylates between 6,000 and 1,000,000 daltons and broad polyacrylamides between 7,950 and 725,000 (PDI 1.8-3.0) were also purchased from American Polymer Standards Corporation. Random polystyrene-co-methylmethacrylates with a molecular weight 123,000-325,000 and the polidispersity (1.8-2.5) were prepared at the Slovak Academy of Sciences, Bratislava (Table 1).

Water Soluble Polymers. Chromatographic measurements were carried out with 0.02 M Na_2SO_4 and methanol as eluents over a polyhydroxymethacrylate sorbent. The gel was packed in a 300-mm stainless column with 8-mm internal diameter. The Shodex OHpak SB-804 HQ column was obtained from Showa-Denko (Tokyo, Japan). The mobile phase flow rate was 0.5 mL/min and 20 μl of a 0.05% wt. aqueous polymer solutions was injected. Polymer samples were injected in a pure solvent (0.02 M Na_2SO_4). All measurements were carried out at ambient temperature. Type I deionized water with a resistivity ≥ 16.7 mΩ-cm (Continental Water, San Antonio, TX, USA) was fileter through a 0.2 mm nylon membrane filter. HPLC grade methanol was purchased from Fisher Scientific (Norcross, GA, USA). The HPLC system was identical to that described for organically soluble polymers.

Cloud Point Measurements. A Bausch and Lomb (New York, NY, USA) Spectronic 20 spectrophotometer operating at 340 nm and ambient temperature was utilized for cloud point measurements. Capped scintillated glass samples vials filled with 2 mL of liquid, were employed. Measurements were performed at a polymer

Table 1. Properties of Random Copolymer Polystyrene-co-methylmethacrylate Samples

Sample	Styrene (Mol %)	$M_w \cdot 10^{-3}$	$M_n \cdot 10^{-3}$	M_w/M_n
1	0	325	181	1.80
2	5.1	318	148	2.15
3	10.2	262	135	1.92
4	20.4	247	106	2.33
5	25.5	200	100	2.00
6	50.7	168	76	2.21
7	75.5	137	65	2.09
8	90.2	155	64	2.40
9	95.1	123	54	2.28
10	100	152	61	2.49

concentration of 1.0 mg/mL. All measurements were carried out at a temperature of 23 ±1 °C. Samples were agitated with magnetic stirring bars.

Results And Discussion

Application of LC-LCS to the Characterization of Random Copolymers. It has been generally speculated, and mathematically predicted [10], that liquid chromatography under critical conditions or at the exclusion-adsorption isotherm could only be applied to the characterization of one block in a diblock polymer, the central portion of a tri-block copolymer or the backbone of a grafted chain. That is, the LC-CAP condition acts to eliminate the influence of the free end on the separation of the molecule. Hence, the copolymer elutes according to its central block or backbone unit. Clearly, such LC-CAP conditions would not be expected to be applicable to the characterization of random copolymers. While this has not been proven experimentally, sequence lengths can be quite large in copolymers synthesized by radical techniques. Therefore, one might anticipate that methods which combine exclusion with enthalpic separations could be applicable for copolymers produced from monomers having relatively different reactivity ratios. In particular, if one would evaluate a copolymer which was rich in one of the monomers where long blocky have been shown to exist by NMR. Given this, the authors of this chapter sought to investigate a common copolymer based on styrene and methylmethacrylate by LC LCS.

Figure 5 shows calibration curves for polymethylmethacrylate in THF/n-hexane at various percentages of the non-solvent. It is clear that the LC LCS condition is not as sensitive to eluent composition as LC-CAP compositions are and one may vary the n-hexane non-solvent level by ±2% without any noticeable influence on the retention volume or the vertical nature of the calibration curve. Figure 6 shows a chromatogram for polystyrene-co-methylmethacrylate in pure THF as well as at the LC-LCA for PMMA. It is important to note that this separation was carried out at the LC LCS for PMMA since this involved a lower n-hexane level (42 vol%) then in the LC LCA for polystyrene (73 vol%). Clearly, only at the LC LCS for PMMA can one identify a condition where one component of the copolymer is eluted in the SEC mode.

Figure 7 shows chromatograms for a series of polystyrene-co-methylmethacrylates of varying compositions at the LC LCA for PMMA (58/42 vol% THF/n-hexane). Shifts in the retention volume and peak area are evident. Figure 8 plots the ratio of peak height obtained under the LC LCS conditions versus the peak height in the pure solvent (THF). Similarly Figure 9 plots the ratio of the peak area at the LC LCS to that in pure THF. Clearly, there is correlation between both peak height (Figure 8) and peak area (Figure 9) with copolymer composition and either can be used as an index of copolymer composition. The authors believe this is the first reported evidence of the use of a coupled (entropic-enthalpic) isocratic separation for the characterization of copolymer composition. Figure 10 illustrates a plot of the

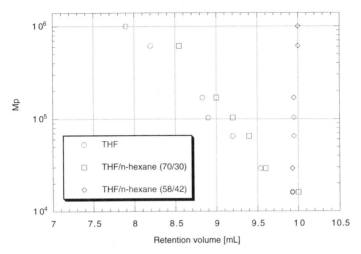

Figure 5. A plot of the molecular weight (daltons) as a function of the retention volume (mL). The calibration curves for narrow polymethylmethacrylate standards in a mixed eluent (THF/n-hexane) are shown at various compositions, expressed as volume percentages.

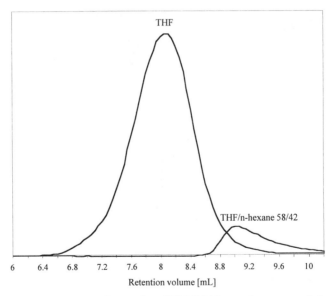

Figure 6. Chromatogram of a random P(S-MMA) copolymer containing 10 mol % styrene in pure THF as well as at the LC-LCA for PMMA.

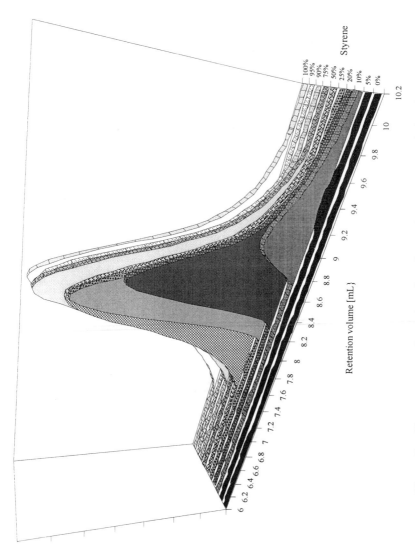

Figure 7. Chromatogram of a series of random polystyrene-co-methylmethacry-lates separated at the LC LCA for PMMA (58 vol% THF/42 vol% n-hexane).

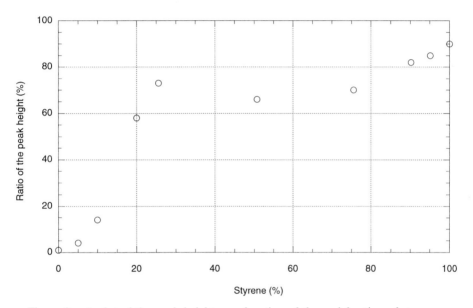

Figure 8. A plot of the peak height as a function of the mol fraction of styrene in a polystyrene-co-methylmethacrylate polymer. Measurements were performed at the LC LCA for (42 vol% n-hexane in THF/n-hexane).

212

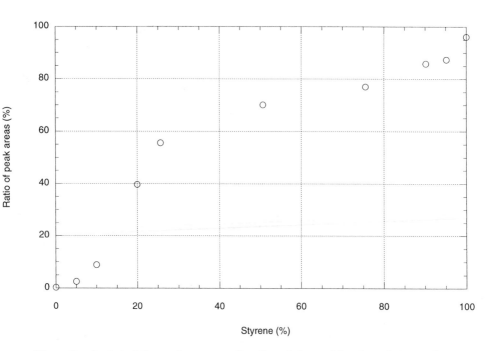

Figure 9. A plot of the peak area as a function of the mol fraction of styrene in a polystyrene-co-methylemthacrylate polymer. Measurements were performed at the LC LCA for PMMA (42 vol% n-hexane in THF/n-hexane).

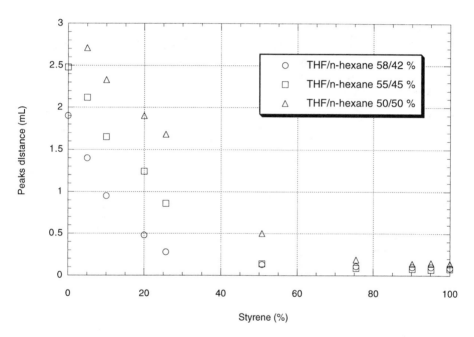

Figure 10. A plot of the retention volume as a function of the styrene content of a copolymer. Measurements were performed at the LC LCA for PMMA (42 vol% n-hexane in THF/n-hexane) as well as at 45 and 50 vol% n-hexane in a THF/n-hexane mixture)

retention volume as a function of the styrene content of a copolymer. Increasing the non-solvent (n-hexane) level has the effect of reducing the slope of the curve, providing a larger range over which the copolymer composition can be estimated (up to 60% styrene). Since many engineering materials, including styrene-butadiene rubber, are based on 3:1 ratios of glassy to amorphous monomers, well within the range of analysis of the method proposed herein, the LC LCS method can be viewed as a practical tool for polymer characterization. Furthermore, by separating first according to size, and then providing an either off-line or on-line (using a SEC column in sequence to the LC LCS column) fractionation, the composition and molecular size distributions can be decoupled. This has been demonstrated in reference [13].

Future Work. The past two decades have seen a, first gradual, then recently more enthusiastic study of coupled entropic-enthalpic liquid chromatography processes. The methodologies have been shown to be effective in the characterization of functional groups on oligomers, for block copolymer, blends, as well as in the characterization of tacticity and, with the results presented herein, for high molecular weight random copolymers. While these advances are intriguing, and the number of reports of molecular weight independent exclusion has become quite large [6,7], the future of these methodologies lies in there ability to resolve chromatographic problems. The coming two years should determine if this method is an experimental curiosity whose mechanism is becoming understood [9] or a tool to, for example, deconvolute various distributions which are normally superimposed and render the interpretation of LC results problematic.

Acknowledgments
The authors would like to thank Dr. Dusan Berek from the Slovak Academy of Sciences in Bratislava for the random P(S-co-MMA) samples.

Literature Cited

1. Belenkii, B.G., Gankina, E.S., Tennikov, M.B., Vilenchik, L.Z., Dokl. Akad. Nauk SSSR, **1976**, *231*,1147 and J. Chromatogr., **1978**, *147*,99.
2. Berek, D., Makromol. Symp., **1996**, 110,33.
3. Hunkeler, D., Janco, M., Guryanova, V.V., Berek, D., in *Chromatographic Characterization of Polymers*, Provder, T., Barth, H., Urban, M., Eds., ACS Books, Washington, DC, **1995**.
4. Bartkowiak, A., Hunkeler, D., Berek, D., Spychaj, T., "Novel Methods for the Characterization of Water Soluble Polymers", in press.
5. Berek, D., in *Chromatography of Polymers: Charactrization by SEC, FFF and Related Methods for Polymer Analysis"*, Provder, T., ed., ACS Books, Washington, DC, **1998**.
6. Hunkeler, D., Janco, M., Berek, D., in *Strategies in Size Exclusion Chromatography*, Dubin, P., Potschka, M., eds., ACS Books, Washington, DC, **1996**.
7. Pasch, H., Advances in Polymer Science., **1997**, 128,1.

8. Hunkeler, D., Janco, M., Guryanova, V.V., Berek, D., in *Chromatographic Characterization of Polymers,* Provder, T., Barth, H., Urban, M., Eds., ACS Books, Washington, DC, **1995**.

9. A. Bartkowiak, R. Murgasova, M. Janco, D. Berek, D. Hunkeler, T. Spychaj, "Mechanism Of Liquid Chromatography Of Macromolecules Limiting Conditions Of Adsorption", in press.

10. Skvortsov, A.M., Gorbunov, A.A., *J. Chromatogr.,* **1990**, 507,487.

11. Entelis, S.G., Evreinov, V.V., Gorshkov, A.V., Adv. Polym. Sci. **1987**, *76*, 129

12. Berek, D., Janco, M., Kitayama, T., Hatada, K., *Polym. Bull.*, **1994**, *32*, 629.

13. Bartkowiak, A., Hunkeler, D., "Liquid Chromatography under Limiting Conditions: A New Method for Copolymer Characterization", in press.

Chapter 15

Full Adsorption–Desorption/SEC Coupling in Characterization of Complex Polymers

Son Hoai Nguyen and Dušan Berek

Polymer Institute of the Slovak Academy of Sciences,
842 36 Bratislava, Slovakia

A novel liquid chromatographic method is presented that allows molecular characterization of multicomponent polymer systems consisted of chemically different constituents. Polymer sample under analysis is adsorbed onto an appropriate adsorbent. Particular constituents are then successively released from adsorbent by the controlled desorption using e.g. a stepwise eluent gradient. This results in the fractionation of sample, predominantly according to its chemical composition. The fractions are directed into an on-line liquid chromatographic (size exclusion or liquid adsorption chromatographic) system for further separation which provides data on their mean molar masses and molar mass distributions. The above idea was applied to several multicomponent polymer blends and precise values of the molar masses and molar mass distributions of all blend constituents could be assessed. Promising results were obtained also in the case of block and random copolymers.

Tailored polymeric materials often comprise components with different chemical compositions. We speak about complex polymer systems, e.g. blends, copolymers, sequenced and functionalized polymers. The direct molecular characterization of complex polymers by size exclusion chromatography is often not enough precise since the size of macromolecules in solution depends not only on their molar mass and on the thermodynamic quality of solvent but also on polymer chemical and physical structure. Consequently, the retention volumes of copolymers reflect also their chemical composition, lengths of blocks, etc. In the case of polymer blends, their components must be independently detected and the mutual effects of coeluting macromolecules with different nature must be suppressed. Therefore, the combinations of different separation mechanisms are to be applied in liquid chromatography (LC) of complex polymers and we speak about two-dimensional or coupled LC procedures (1,2). One of these combined methods is called full adsorption - desorption / liquid chromatography (FAD/LC) coupling (3).

FAD/LC consists of a series of independent steps. In the first step, all n or at least n-1 constituents of the complex polymer are fully retained within an appropriate sorbent. Sorbent that exhibits different affinity to macromolecules in dependence on their chemical nature is packed into a (mini)column that is called Full Adsorption - Desorption (FAD) column. Nonadsorbed macromolecules passing FAD column are directly transported into an appropriate on-line LC column for molecular characterization. After the LC analysis has been completed, operational conditions (eluent composition, temperature, pressure) are suddenly changed so that a fraction is released from the FAD column to be evaluated in the LC system. The displacement and LC evaluation steps are repeated until polymer sample is completely characterized.

The described full adsorption - desorption procedure resembles solid phase extraction (SPE) techniques that are often used in the analysis of low molecular analytes. However, the adsorption of macromolecules can be easier controlled provided a well designed system adsorbing liquid - adsorbent is applied. Consequently, recovery and repeatability in the FAD of macromolecules is substantially higher than in the SPE of small molecules. Various liquid chromatographic, spectrometric etc. procedures can be combined with full adsorption - desorption steps, however, size exclusion chromatography (SEC) is the method of choice for high polymers and one speaks about an FAD/SEC coupling.

We have recently shown (3-5) that precise and repeatable mean molar mass and molar mass distribution values can be obtained for constituents of model mixtures of two or three chemically different homopolymers after appropriate optimization of the FAD column packing, and both the polymer adsorption promoting liquid (ADSORLI) and the polymer desorbing liquid (DESORLI). The packing of the FAD column was preferably nonporous bare silica gel to suppress the polymer diffusion effects that may cause excessive broadening and even splitting of the SEC peaks of the retained/desorbed polymer (4). The size of the FAD column had to be just large enough to exhibit appropriate sample capacity, i.e. to retain enough polymer for subsequent size exclusion chromatographic characterization. In turn, the ADSORLI had to be strong enough to ensure the quantitative entrapment of selected sample constituent(s) but not to promote their adsorptive retention within SEC column (4).

In the present study we tested the performance of the FAD/SEC method in discrimination of some selected complex polymer systems and extended its use to copolymers and quaternary polymer blends of chemically similar constituents.

Experimental

The measuring assembly is schematically shown in Figure 1. Sample solutions were injected into the FAD column by means of valve V1. The adsorptive constituents of the analysed sample were retained within the FAD column packing while the non-adsorptive constituent passed through FAD column unretained and it was directed into the SEC column for characterization. Next, SEC column was equilibrated with a new eluent introduced via valve V2 and possessing increased desorbing power. Subsequently, valve V3 was switched so that this new displacing eluent (e.g. single DESORLI or an appropriate solvent mixture) was directed into FAD column to release a selected constituent of polymer blend or a copolymer fraction and to transport it into the SEC column. The chosen displacer or displacer set was prepared in extra containers or, alternatively, it was produced with help of the Knauer (Berlin, Germany) gradient making device. The latter approach was very convenient for separation of multicomponent polymer systems. The resulting displacer composition had to be, however, smoothed in latter case to avoid the local variations in the displacer composition. Otherwise, polymer recovery could not be effectively controlled. The SEC column equilibration and the FAD column desorption steps were several times repeated with different displacers until whole sample has been processed. Instead of stepwise gradient, also continuous gradient of eluent composition or, alternatively, temperature variations can be used for successive desorption of macromolecules from the FAD column. It is important that the SEC column withstands repeated switching of eluents with different polarities and that the adsorptive properties of the SEC column are as low as possible.

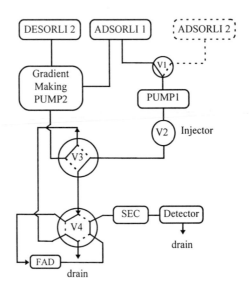

Figure 1. Schematic representation of FAD/SEC assembly. See text for detailed explanation

To optimize experimental conditions for polymer displacement, desorbing properties of various ADSORLI-DESORLI displacers must be assessed. This can be done by a series of independent measurements using the same experimental assembly (Figure 1). If necessary, FAD columns of different sizes and packed with different adsorbents can be utilized. In present work, the two-component displacers of various compositions and displacing strengths were introduced into the FAD column containing known preadsorbed amount of a single polymer under study. Both amount and molar mass characteristics of desorbed macromolecules were measured by means of the SEC column and detector. After the desorption experiment had been completed the FAD column was flushed with the pure DESORLI to remove the rests of polymer. DESORLI was then displaced by the pure ADSORLI, the initial amount of polymer was again applied onto FAD column and the desorption procedure was repeated with another displacer mixture. In this way, the plots of desorbed polymer amount vs. displacer composition were constructed. Such dependences are called the dynamic integral desorption isotherms.

The common HPLC pumps and valves were used. The detectors were either an evaporative light scattering detector ELSD Model DDL-21 (Eurosep, Cergy St. Christophe, France) or UV variable wavelength photometers (Knauer) operating at wavelength 233 nm for mixed eluents dichloroethane / tetrahydrofuran. Since UV detector response at 233 nm was found to be the same for both PS and PMMA, it is linearly related to P(S-MMA) copolymer amount. In any case, the desorbed amounts of (co)polymers were calculated from detector response employing appropriate calibration for the same experimental conditions.

A series of model homopolymers and their mixtures was investigated: PS, PMMA, poly(n-butylmethacrylate)s (PBMA) and poly(ethylene oxide)s (PEO) with different molar masses. They were obtained from Pressure Chemicals (Pittsburgh, PA, USA), Polymer Laboratories Co. (Church Stretton, UK), Polymer Standards Service Co. (Mainz, Germany) and TOSO Co. (Shinnanyo, Japan).

The selected random P(S-MMA) (from Prof. S. Mori, Mie University, Tsu, Japan) (Table I) and block P(MMA-b-GMA) copolymers (from Dr. G. Hild, Institute Sadron, Strasbourg, France) were also subject to preliminary investigations. They were prepared by radical (6) or anionic (7,8) copolymerization, respectively.

**Table I. Styrene content (6) and molar mass
characteristics of P(S-co-MMA) random copolymers
(using PS standards calibration)**

Sample	Styrene (mol%)	$Mw.10^{-3}$	$Mn.10^{-3}$	Mw/Mn
I	85.5	174	89.8	1.94
IV	57.4	115	57	2.02
VIII	26.5	148	65.9	2.24

ADSORLI solvents were toluene and dichloroethane (DCE) for polar polymers and dimethylformamide (DMF) for PS. Tetrahydrofuran (THF) was used as a typical DESORLI. All solvents of analytical grade were distilled prior to use. DCE was stabilized with 50 ppm amylene.

The packings of the FAD columns were nonporous silicas (8 μm in diameter) either bare or C18 bonded. Nonporous silica was prepared in this laboratory by agglutination of a highly pure spheroidal mesoporous silica gel Silpearl (Kavalier

220

Votice, Czech Republic) for 2 hours at 1200°C. After rehydroxylation, this material was bonded with dimethyl octadecyl groups by Prof. B. Buszewski (Corpernicus University, Torun, Poland) (9).

Results and discussion

The dynamic integral desorption isotherms for selected homopolymers in system ADSORLI (toluene) and DESORLI (THF) are shown in Figure 2. The effect of polymer chemical nature on the course of dynamic desorption isotherms is evidenced. It is clearly seen that PBMA can be easily separated from PMMA in the present system of adsorbent-ADSORLI-DESORLI. On the other hand, PEO remained fully adsorbed within FAD column in pure THF. To release PEO from the bare silica sorbent, DMF can be used. Typical examples showing the performance of the FAD/SEC coupling in the separation and molecular characterization of various polymer blends are given in Table II. The molar mass data obtained by direct SEC analysis of single polymers without and with FAD step agree well. It is to be concluded that FAD procedure can readily discriminate medium polar and polar polymers, even if they are chemically rather similar. The separation of nonpolar polymers, e.g. polystyrenes was found to be more complicated. The adsorptive properties of nonpolar macromolecules are not well pronounced using bare silica FAD column packing. On the other hand, polystyrenes could be quantitatively retained on the silica C_{18} particles in dimethylformamide as an ADSORLI if their molar mass exceeded 90,000 g/mol. In this system, polar polymers were not retained and could be directly forwarded into SEC column. The optimization of this „reverse FAD system" is subject to our further study.

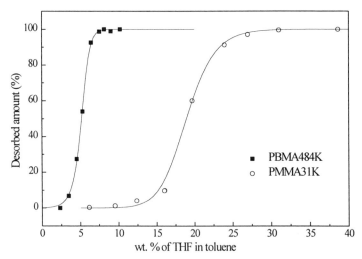

Figure 2. Dynamic integral desorption isotherms for homopolymers with M_W values as indicated. System nonporous silica / toluene / THF. FAD column (45x2mm); preadsorbed polymer amount 0.01mg. ELS detector.

Table II. Examples of molar mass characteristics of single polymers and polymers in blends separated and characterized by FAD/SEC coupling. Injected amount was 0.005mg for each polymer.

Procedure	Polymer	Single or mixed displacer	$M_w \cdot 10^{-3}$ (g/mol)	$M_n \cdot 10^{-3}$ (g/mol)	M_w/M_n
SEC for single polymers	PS350K	toluene	336	179	1.88
	PBMA484K	THF	499	406	1.23
	PMMA31K	THF	30.6	22.3	1.37
	PMMA461K	THF	446	249	1.79
	PEO45K	DMF	43.4	35.9	1.21
FAD/SEC for blend components	PS350K	Toluene	329	179	1.85
	PBMA484K	8 wt.% THF in toluene	507	409	1.24
	PMMA31K	THF	29.4	22.3	1.32
	PBMA484K	8 wt.% THF in toluene	493	391	1.26
	PMMA461K	THF	429	313	1.37
	PEO45K	DMF	44.0	35.5	1.24
	PS350K	Toluene	356	188	1.89
	PBMA484K	8 wt.% THF in toluene	502	415	1.21
	PMMA31K	THF	28.6	21.0	1.36
	PEO45K	DMF	42.6	34.9	1.22

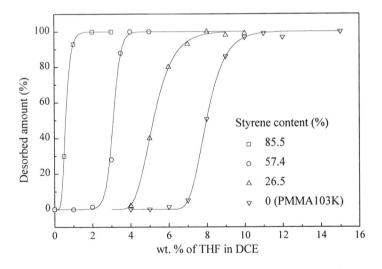

Figure 3. Dynamic integral desorption isotherms for random copolymers P(S-co-MMA) (Table I) and for PMMA103K for comparison in the system nonporous silica / DCE / THF. FAD column (150x3.3mm); preadsorbed polymer amount 0.015mg. UV detectors at 233 and 264nm.

The dynamic desorption isotherms for selected P(S-co-MMA) random copolymers and for some PGMA-b-PMMA-b-PGMA triblock copolymers are shown in Figures 3 and 4, respectively. Positions of the dynamic desorption isotherms strongly depended on the copolymer chemical structure. In the case of random copolymers, differences in their composition necessitated remarkable changes in the strength of displacer that was needed to desorb macromolecules (Figure 3). In the case of triblock copolymers, the end blocks seemed to play more important role in the adsorption-desorption processes than did the central blocks (Figure 4). The preliminary results indicate potential of the full adsorption-desorption procedure to discriminate copolymers according to their composition. The subsequent SEC separation step can produce at least semi-quantitative information on the molar mass distribution of each fraction.

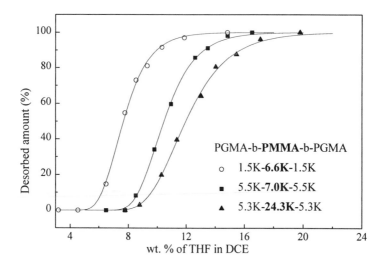

Figure 4. Dynamic integral desorption isotherms for triblock copolymers P(GMA-MMA-GMA) with block lengths (M_n) as indicated (7,8). System nonporous silica / DCE / THF. FAD column (30x3.3mm); preadsorbed amount 0.015mg. ELS detector.

Conclusions

Full adsorption-desorption / SEC (FAD/SEC) coupling allows discrimination and molecular characterization of various complex polymers. In the case of polymer blends, also chemically rather similar constituents can be readily separated. In contrast to other LC techniques suitable to discrimination of two- and multi-component polymer blends (1,2,12), such as liquid chromatography at the critical adsorption point, gradient polymer elution chromatography or liquid adsorption chromatography, FAD/SEC combination enables also precise and repeatable determination of molar masses and other molecular characteristics of separated blend constituents. The utilization of the FAD/SEC method for molecular characterization of copolymers is limited to the systems in which constituents exhibit rather different adsorptive properties, e.g. due to different polarities. The course of polymer adsorption/desorption often depends not only on the chemical nature but also on the molar mass of macromolecules, as well as on their load on the adsorbent and on the presence of further analytes (5). Therefore, appropriate adsorbent-ADSORLI-DESORLI systems must be identified to take provision for the above effects. Other separation methods can be combined with the full adsorption-desorption procedure, as well. For example, the FAD fractions can be characterized by liquid adsorption chromatography or by mass spectrometry. In this way, the FAD technique represents an important step in the multidimensional liquid chromatographic and spectrometric characterization of complex polymer systems. Full adsorption-desorption procedure can be used also for reconcentration of highly diluted polymer solutions (10,11). Further developments include preparation of tailored FAD column packings and identification of highly selective displacers. It is anticipated that desorption of macromolecules from an adsorbent can be controlled not only by changing composition of displacers but also by temperature and possibly also by pressure variations, both stepwise or in the continuous gradient mode.

Acknowledgement

This work was supported by the Slovak grant agency (VEGA project No. 2/4012/97) and the US-SK Sci. tech. cooperation, project 007/95. The authors are indebted to Prof. S. Mori and Dr. G. Hild for samples of copolymers and Prof. B. Buszewski for bonding silica sorbent.

Literature Cited

1.   Glöckner, G. *Gradient HPLC of Copolymers and Chromatographic Cross-Fractionation*, Springer-Verlag , Berlin 1991.
2.   Berek, D. *Molecular Characterization of Complex Polymers by Coupled Chromatographic Procedures*, This book, T. Provder, Ed., ACS Press, Washington, D.C. 1998.
3.   Jančo, M., Prudskova, T., Berek, D. *Polymer*, **1995**, *36*, 3295.
4.   Jančo, M., Prudskova, T., Berek, D. *Int. J. Polym. Anal. Charact.* **1997**, *3*, 319.
5.   Nguyen, S.H., Berek, D. *Chromatographia*, in press.
6.   Mori, S., Uno, Y. and Suzuki, M. *Anal. Chem.* **1986**, *58*, 303.
7.   Hild, G., Lamps, J-P., Rempp, P. *Polymer* **1993**, *34*, 2875.
8.   Hild. G., Lamps, J-P. *Polymer* **1995**, *36*, 4841.
9.   Nondek, L., Buszewski, B. and Berek, D. *J. Chromatogr.* **1986**, *367*, 171.
10. Nguyen, S.H., Berek, D., Chiantore, O. *Polymer*, in press.
11. Nguyen, S.H., Berek, D. *J. Polym. Sci., Part A: Polym. Chem.*, accepted.
12. Kilz, P., Krüger, R.P., Much, H., Schulz, G., In *Chromatography of Polymers: Characterization by SEC and FFF*; Provder, T., Ed., ACS Symp. Series 247; ACS Books: Washington DC, 1995, p. 223.

POLYMER APPLICATIONS

Chapter 16

Size Exclusion Chromatography–FTIR Analysis of Polyethylene

James N. Willis[1], James L. Dwyer[1], Xiaojun Liu[1], and William A. Dark[2]

[1]Lab Connections, Inc., 201 Forest Street, Marlborough, MA 01752
[2]Waters Corporation, 34 Maple Street, Milford, MA 01757–3696

Infrared spectroscopy combined with size exclusion chromatography has been used to study short chain branching in a series of high molecular weight polyolefins. The polymers were separated using traditional SEC techniques and the eluted samples collected using an off-line solvent removal device that provides solvent free samples that are compatible with FTIR instrumentation. Normal sampling methods and concentration levels were used in both the size exclusion and FTIR portion of the experiments. Results indicate that FTIR measurements can be used to determine branching distribution over a wide molecular weight range for high and low density polyethylene samples. Mathematical algorithms have been developed which allow determination of ethyl, butyl, and hexyl branches as a function of the molecular weight distribution of the separated polymer. Descriptions of the advantages and limitations of the technique are discussed.

Ethylene-based copolymers account for over 50 billion pounds of products per year in the United States alone. The performance of the final products made from these copolymers is dramatically effected by the inclusion of a small number of short chain branches along the backbone of the polyethylene chain. The measurement and characterization of short-chain branching is possible using FTIR techniques, however, without information about where in the molecular weight distribution the branches occur, the results are not very useful. A method of combining size exclusion chromatography (SEC) and FTIR spectroscopy has been developed and applied in the characterization of short chain branches in polyolefins.

Experimental

Samples of commercial grade high and low-density polyethylene were weighed and dissolved in trichlorobenzene (TCB) at concentrations of 0.15% (wt./vol.). Samples of each material were heated to 145 ^{0}C to insure dissolution. They were then injected onto a Model 150C Plus high temperature chromatograph (Waters Corp., Milford, MA.). A bank of three mixed-bed Styragel HT columns (Waters Corp., Milford, MA.) were used to separate the polymers. Calibration of the columns was carried out by injecting a mixture of four narrow dispersed polystyrene standards, collecting them, and measuring the location of the four peaks on the infrared collection discs. The samples were collected using a Model 310HT Collection Module (Lab Connections, Inc., Marlborough, MA.). A Model 510P FTIR spectrometer equipped with a broad band MCTB detector was used to collect the IR spectra (Nicolet Corp., Madison, WI.). The IR spectra were obtained at 4 cm-1 resolution and each data set consisted of 100 coadded spectra. Once collected, the IR data were manipulated using GRAMS/32 (Galactic Industries, Salem, NH.) and 3D/IR software (Lab Connections, Inc., Marlborough, MA.) to extract the desired information relating to peak area, band intensity, and ratios of bands. Further experimental details have been published elsewhere (Ref. 1).

If an antioxidant is used in the mobile phase it must be removed before any IR results are attempted. This was accomplished by washing the sample with methanol or methylene chloride. Typical antioxidants are soluble in these solvents and therefore are removed while the sample remains on the sample disc. Figure 1 shows the IR spectrum of a sample before and after washing.

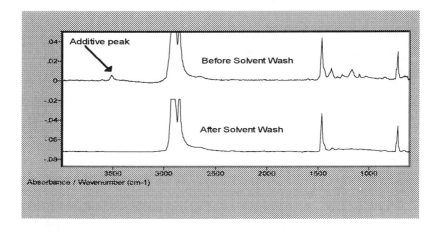

Figure 1

Method

It is necessary to know what co-monomers were used in the preparation of the sample in order to use this method. If this information is known, a peak associated with the monomer of interest is then measured. The band assignments of the frequencies associated with monomers in a number of substituted linear low density polyethylene samples has been reported by Maddams and Woolington (Ref. 2), Blitz and McFaddin (Ref. 3), and Blitz

(Ref. 4). These assignments were used in this work. Absorption bands associated with side chains of ethyl, butyl, hexyl, and isopropyl are located at 770, 894, 889, and 920 cm^{-1}, respectively. The molar absorptivities of these bands are very low and require multiple IR scans in order to improve the signal to noise ratio of each measurement. Once the band is selected, a narrow spectral range associated with that band is chosen and the area integrated. A reference band within the co-polymer is also required to correct for the variation in thickness in the polymer film on the sample disc. An overtone band found at 4321 cm^{-1} is chosen as the internal reference. Because the overtone arises from combinations of several fundamental vibrational modes, it is independent of any single vibrational mode. It would have been possible to use other bands for an internal reference, for example, the total absorption associated with the C-H region. The overtone band was chosen because it is closer in intensity to the bands arising from the short chain branches.

In setting up the method, first, an average value of the number of short chain branches, SCB, per 1000 CH_2 in the sample was obtained from NMR measurements. Next, an average IR spectrum of the polymer is obtained by coadding the spectra of the entire chromatographic deposit. From this average spectrum the area of a band know to arise from the side chain, A_{scb}, and that of the internal reference peak, A_{ref}, are measured. From this data the average value from the NMR and IR measurement can be related through a constant, **K,** in Equation 1.

$$\%SCB = [(A_{scb})/(A_{ref})] * K \qquad (1)$$

Following this measurement, chromatography is preformed on the sample, the sample collected, and a series of IR spectra is obtained from high to low molecular weight. Finally, Equation (1) is applied to each IR spectrum obtained along the distribution curve of the polymer.

The data are smoothed with a Savitsky/Golay 2nd order polynomial with 15 smoothing points using GRAMS/32 software.

Results

The reaction of ethylene and butene results in a co-polymer containing a backbone of CH_2 groups with a small number of ethyl side chains. An example of a typical spectrum of an ethylene-butene copolymer is shown in Figure 2. The inserts on the right and left of the spectrum shows a band at 770 cm^{-1} from the ethyl rocking vibration and a band at 4321 cm^{-1} which is the overtone band, respectively.

Figure 3 illustrates the application of Equation 1 to an ethylene-butene polymer with a reported average ethyl branch content from NMR of 19.3 ethyls/1000 CH_2 groups. The **K** value for this sample was calculated to be 254. The curve labeled "Total CH CGM" is obtained by integrating the IR absorption of all the bands between 2800 and 3100 cm^{-1} and plotting the result vs. molecular weight. The Total CH CGM represents the amount or concentration of sample on the collection disc for any one point in the

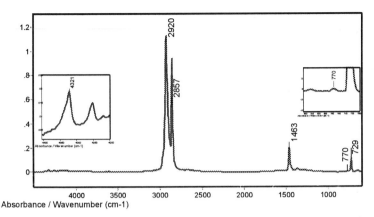

Figure 2

molecular weight distribution. Overlaid on this plot is the calculated distribution of ethyl branches per thousand CH_2 groups. In this sample the ethyl band was integrated from 780 to 760 cm^{-1} and the overtone from 4360 to 4310 cm^{-1}. Using the equation 2:

$$\%Ethyls = [\ A_{770}/A_{4312}]* 254 \qquad (2)$$

The ethyl branch distribution increases from approximately 11 Ethyls/1000 at 645,000 Daltons to 28 at 50,000 Daltons. The data obtained in the main portion of the sample deposit, 645K Daltons to approximately 90K Daltons, appears to be in general agreement with the average value of 19 Ethyls/1000 reported for the unseparated material. The branching content increases significantly at the low molecular weight end of the distribution. The amount of sample is very small at the ends of the deposit making it difficult to accurately measure the intensity of the IR bands.

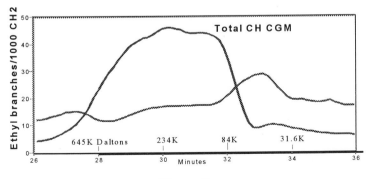

Figure 3

Two additional samples with different levels of SCB were measured using this method. The results are show in Figures 4 and 5. In Figure 4, the average value measured by NMR was 25.2 Ethlys/1000 and in Figure 5 the value was 12 Ethyls/1000. In both cases, the distribution of the data falls within the expected range. In both cases the branching distribution varies considerable from high to low molecular weight.

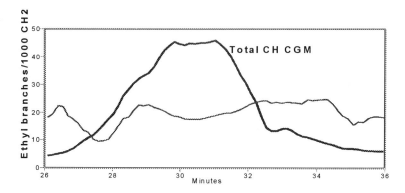

Figure 4

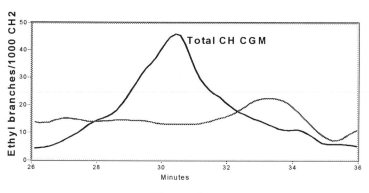

Figure 5

In another sample of a similar material the agreement was not as good. In this case, the reported NMR average value was 20 Ethyls/1000. Using a K value of 254 produced an ethyl branch distribution curve that was low by almost a factor of 2 from the expected value. This data is shown in Figure 6. It is not clear why this sample was different. A subsequent experiment is required to understand this discrepancy. Even considering the apparent inaccuracy in the quantitative method, the trend in the branching distribution may in itself be of considerable interest to the polymer chemist.

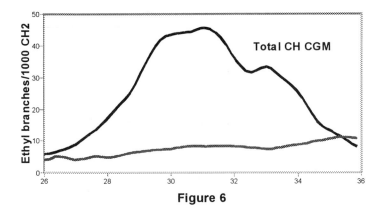

Figure 6

The method has been applied to other samples in which butyl, hexyl and isopropyl side chains are present. The agreement with the reported NMR value is generally good. This work will be reported in a later publication.

Conclusions

The use of SEC combined with FTIR seems to offer a simple method for the determination of short chain branching in copolymers of ethylene. The approach requires no modification in the SEC experimental conditions and the IR data may be obtained using conventional FTIR equipment. It has been applied in a series of ethylene-butene copolymers with general success. It is necessary to remove all additives from the polymer before attempting to determine the branching content. In one case, the method failed. It is not obvious from this work the reason for the failure. More experiments are required in order to determine the origin of this problem.

References

1. J.B.P. Soares, R.F. Abbott, J.N. Willis, and X. Liu, Macromole. Chem. Phys., *197*, 3383-3396 (1996)

2. W.F. Maddams and J. Woolmington, Makromol. Chem., *186*, 1665 (1985)

3. J. P. Blitz and D.C. McFaddin, J. Polym. Sci., *51*, 13(1984)

4. J.P. Blitz, Proc. Am. Chem. Soc., Div. Poly. Mat., Sci. and Eng., *75*, Orlando, FL. Fall Meeting, 1996.

Chapter 17

Factors Affecting Molecular Weight and Branching Analysis of Metallocene Catalyzed Polyolefins Using On-Line GPC with Light Scattering, and Viscometry Detection

Trevor Havard and Peter Wallace

Precision Detectors, Inc., 10 Forge Park, Franklin, MA 02038

Introduction

Metallocene polyolefins (1) are a class of polymers based on a new generation of catalyst. These reactions produce polyolefins and copolymers with controlled co-monomer composition, narrow poly-dispersity and long chain branching.
Metallocene polyolefins provide unique physical properties. Their branching analysis presents an interesting challenge for both viscometry and light scattering detection.
This paper will describe the performance characteristics of a new type of high temperature light scattering instrument that provides three important capabilities: (1) very high sensitivity detector with stable baselines, using a fully optimized temperature controlled optical collection system, (2) two angle measurements at 15° and 90° providing accurate calculations across a very broad molecular weight range, as well as Rg information, and (3) installation inside a high temperature GPC system, reducing inter-detector band broadening and preventing precipitation in the cell.
The measurements obtained using this type of light scattering detector will be compared to the results obtained using a Viscotek model 100 multi-capillary viscometer.

Experimental

The Viscometry /SEC experiments were carried out at Jordi Associates, Bellingham, Massachusetts using two Waters 150C units. The first was equipped with a Viscotek 100 and the second was equipped with a Precision Detectors PD2040 light scattering system. Both systems were run at 145°C with 1,2,4-Trichlorobenzene (TCB) using a 500-mm long Jordi mixed bed divinyl benzene gel.
Other experimental work using the PL GPC-210 was carried out at Polymer Laboratories, Amherst, Massachusetts using the PL GPC-210 equipped with a Precision Detectors PD2040 and two different sets of columns consisting of four mixed B columns at 135°C and four mixed B Columns at 150°C. TCB was the solvent used on both occasions.

The metallocene catalyzed samples examined were Dow Affinity PL1840, and Affinity PL1880, both ethylene-octene copolymerized polyolefins, a commercial grade polypropylene obtained from Montell, PROFAX 6801 and an LLDPE linear low density polyethylene Dow Attane 4200 which is a linear low density polyethylene octene copolymer produced using a heterogeneous catalyst.

For the purposes of this paper

Profax 6801	is named polypropylene sample.
Dow Attane	is named linear polyethylene sample.
Dow Affinity 1880	is named metallocene A
Dow Affinity 1840	is named metallocene B

The samples were prepared by two methods, heating at 150 C in the presence of 250 ppm. BHT antioxidant, and on a hot plate with stirring at 150 C using 250 ppm. antioxidant.

The samples were checked every hour by quenching to 100 C to identify if the samples had completely dissolved. In the case of the polypropylene, temperatures of 160 C were always used. The oven method was preferred.

Branching Analysis (The GPC/Viscometry approach)

The use of the GPC/Viscometer for determining accurate molecular weight values is simple for polyolefins in TCB, but the use of this technology to obtain branching information requires a much more comprehensive determination of the parameters that affect GPC/Viscometry analysis.

The branching analysis using a GPC system combined with a light scattering detector or viscometer requires excellent detector stability and minimized flow fluctuations in order to provide reproducible branching analysis.

Branching analysis performed by viscometry requires the establishment of a universal calibration, using a polynomial fit. The assumption is that a polynomial fit represents exactly the separation conditions of the column set. Linearity of the column set is important to avoid error in the hydrodynamic volume calibration curve.

The modern GPC/Viscometer can produce the specific viscosity via *Poiseuille's Law*

$$P = 8LQ\eta/\pi r^4 \qquad \qquad \textbf{Equation (1).}$$

where P is the pressure in the capillary, L is the length of the capillary, Q is the flow rate in the capillary, η is the viscosity if the solution in the capillary, and r is the radius of the capillary. The GPC/Viscometer produces specific viscosity (η_{sp}) from the capillary using the following equations (1) and (2).

$$(Pi - Po) / Po = \eta_{sp} \qquad \qquad \textbf{Equation (2).}$$

where Pi = the pressure of the solvent and sample combined
Po = the pressure of the solvent in the GPC
η_{sp} = the specific viscosity in the capillary

The Viscotek bridge capillary system uses a derivation of the above equations to obtain the specific viscosity η_{sp}. In order to get molecular weight information, the η_{sp} must be converted to intrinsic viscosity slices. Here the concentration at each slice must be accurately determined using the refractometer.

$$\eta_{sp}/c = [\eta] \qquad \text{where } c \rightarrow 0 \qquad \textbf{Equation (3).}$$

In order to obtain the correct (2) intrinsic viscosity [η], the inter-detector volume must be determined precisely to match the slice viscosity to the slice concentration. The intrinsic viscosity and the universal calibration are then used to determine the true molecular weight of the polymer. Then, by plotting the log intrinsic viscosity versus the log molecular weight, a viscosity law plot can be obtained *(Mark-Houwink Equation)*.

$$\text{Log } [\eta] = \text{Log } K + (a)\text{Log } Mv \qquad \textbf{Equation (4).}$$

The (a) exponent is expected to be between 0.69 and 0.72 for a linear random coil in solution. By comparing the intrinsic viscosity of a branched polymer at each interval of log molecular weight with that of a linear polymer, the level of branching can be determined. The concept of viscometry in principle is very simple, but to obtain reliable measurements, the use of these instruments is often complex for the following reason.

The accurate determination of the slice molecular weight and slice intrinsic viscosity is a function of:

1. Precise flow in the capillary *(Lesec effect)* (3)
2. Inter-detector volume (Precise flow everywhere)
3. A well calibrated set of columns with an accurate polynomial fit representative of the column separation
4. Accurately known concentrations
5. Detector band-broadening corrections

The technique works extremely well, but the user must be aware of the potential difficulties of obtaining the inter-detector volume, band-broadening corrections and the effect of splitting the flow or adding massive dampening to the pumping system.

The *Lesec effect*, which is well known and has been documented, must be considered, especially if high concentrations of viscous polymer solutions are injected, as this action can effectively make minor yet significant variations in the flow.
The concentration of the injected solution should be maintained below the C* - the critical concentration of the solution which reduces the Lesec effect. Unfortunately, this in turn reduces the potential signal of the viscometer and the refractometer, making the measurement more difficult.
Before any branching measurements are made a universal calibration must be obtained where the Mark Houwink exponent must have a realistic value of between 0.69 and 0.73. Also a set of broad distribution linear polymers must also be analyzed to determine that the system does not detect branching where there is none. The inter-detector constant can be a major factor in determining the correct results.
The final calculation that enables one to observe the degree of branching as a function of molecular weight is known at g' plot. The measurement is derived from the following equation:

$$g' = [\eta]_{br} / [\eta]_{ln} \qquad \textbf{Equation (5).}$$

g' (4) can now be plotted as a function of log molecular weight and the effect of branching observed clearly as a function log molecular weight. If a GPC/Viscometer is used, all the above parameters are carefully observed, and low concentrations are injected, the system will produce reliable branching information across the majority of the polyolefin distribution.

The Precision Detectors PD2040 High Temperature Light Scattering Platform
The PD2000 detector axial design makes measurements at 90° and 15°.

This has been developed with a version of the optical bench that can be installed inside the Waters 150C or the PL GPC-210. The laser light source is aligned and directed through a light tube where the laser is focused onto the cell (Fig.. 1). The cell is 10 μl in volume with a very small optical volume of less than 0.01 μl (Fig.. 2). This advanced cell design combines a large solid angle at each detector to maximize signal to noise with beam focusing to achieve very high sensitivity. A number of other modifications have been made to the laser assembly to provide exceptional baseline stability. The current diode laser used for this instrument is a 20 milli-watt laser at 680 nm. wavelength, although this is easily interchangeable with other lasers.

Branching Analysis (via Light Scattering)
The use of a light scattering detector to analyze branching can be achieved via four techniques:

1. Comparing absolute molecular weight at each retention volume between the expected branched polymer and the linear polymer of the same chemical composition. This has been reviewed by A. E. Hamielec (5).

2. Developing a log Rg versus log molecular weight relationship similar to the viscosity plot. There are only a few references in the literature for Polyolefins. Benoit et al. (6) appears to be the first to record any Rg data for Polyolefins.

3. Comparing hydrodynamic radius of the branched polymer to that of the linear polymer (Dynamic light scattering detector). There is not any currently published data that has been found for Polyolefins although (7) we generated data for polystyrenes in flow mode using GPC/SEC.

4. Comparing the hydrodynamic radius to the radius of gyration. This work has yet to be published by anyone using size exclusion chromatography.

The PD2000 optical bench is designed in principle to provide all four capabilities. This paper will focus on the second method, which is the use of a light scattering detector to analyze the polyolefin using accurately determined molecular weights and radius of gyrations. The other approaches will be a subject of further papers.

To obtain a radius of gyration, more than one angle must be measured. Rg can be obtained via either the Zimm Plot (8) or the dissymmetry method. The PD2040 currently uses the dissymmetry method (9,10) via the Debye (11) particle function which appears to fit exceptionally well for random coils and long chain branching. The Rg values calculated using GPC/PD2040 can be compared with the intrinsic viscosity values from GPC/Viscometry method for obtaining branching information.
Rg is the z average Radius of Gyration and therefore is very sensitive to the high molecular weight portion of the distribution. Rg is calculated independently of concentration and therefore is not subject to the effect of the inter-detector volume. The absolute molecular weight is obtained directly by the relation of excess scattered light at the low angle and is not subject to the effects of the errors in polynomial fits for the hydrodynamic volume.

236

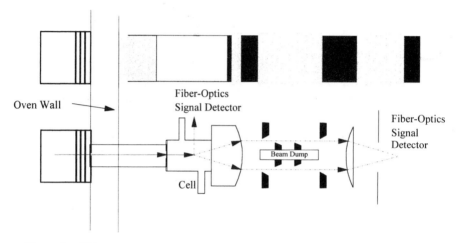

Figure 1. This type of light scattering design enables the laser to be accommodated outside the GPC system; the optics are placed inside the high temperature oven.

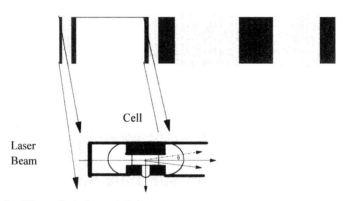

Figure 2. The cell design minimizes band broadening while maximizing signal collection; all surfaces remain hot preventing sample precipitation.

Using the Combination of a Refractometer and PD2000

The mathematics that describes the relationship between molecular weight and scattered light has been well-established (6,7,8,9,10,11). Light scattering intensities of the 90° and 15° angle produce the following equations:

$$Ls_{(90°)} = K_{(Ls90°)} \, M_w \, c \, (dn/dc)^2 P(90°) \qquad \text{Equation (6).}$$

$$Ls_{(15°)} = K_{(Ls15°)} \, M_w \, c \, n \, (dn/dc)^2 P(15°) \qquad \text{Equation (7).}$$

where:
Ls = the excess Raleigh Light Scattering Signal
K = the optical constant for the detector
dn/dc = the change in refractive index as the concentration changes
$P(\theta)$ = the ratio of scattered intensities at angle θ to that of zero degrees
c = concentration
n = solvent refractive index.

In the case of the 90° detector, the intensity of light is independent of the refractive index of the solution (9). This is an advantage when using the PD2000/RI detector design, because it provides an opportunity to change from one solvent system to another, while maintaining constant detector parameters.

The incorporation of the light scattering detector into a temperature-controlled oven with a refractometer improves the accuracy and precision of the measurements in the following ways:

1. Minimized and constant inter-detector volume.
2. Sensitivity and stability in a temperature-controlled oven.

In order to use the light scattering detector successfully, the refractometer is used to calculate dn/dc and concentration. The equation that describes the use of a refractometer for dn/dc and concentration slice calculations is as follows:

$$RI_{(sig)} = K_{(RI)} \, c \, dn/dc \qquad \text{Equation (8).}$$

By dividing equation (6) by (8), a new relationship for the dual detector can be derived which enables the detector with the accompanying software algorithms to become a true absolute detector independent of the SEC system.

$$\frac{Ls_{(90°)}}{RI_{(sig)}} = \frac{K_{(Ls90°)} \, M_w \, (dn/dc) \, P(90°)}{K_{(RI)}} \qquad \text{Equation (9).}$$

A single well characterized low molecular weight standard with a known dn/dc can be used to calibrate the optical constant $K_{(Ls90)}/K_{(RI)}$.

$$K(High) = \frac{K_{(Ls15°)}}{K(RI)} \qquad \text{Equation (10).}$$

Macromolecules with molecular sizes below 12 nm. produce little angular dissymmetry between the 15° and 90° detectors. This can be seen in (Fig..3) for angular measurements predicted by the Debye function for random coils. Any monodisperse standard can be used to calibrate the 90° detector.

$$\frac{Ls_{(15')}}{RI_{(sig)}} = \frac{K_{(Ls15')} M_w\, n\, (dn/dc)\, P(15°)}{K_{(RI)}} \qquad \text{Equation (11).}$$

A similar equation can be derived for the 15° detector that includes the refractive index (n) of the solvent. Therefore, we now have three constants and the inter-detector volume to consider to fully calibrate the PD2000/RI system:

K(Low) 15°	Optical constant for the RI and 15°
K(High) 90°	Optical constant for the RI and 90°
K(RI)	Optical constant for the RI alone
Inter-detector Volume	Volume between light scattering and RI detector cells

Determination of Rg Using Two Angles

Large molecules scatter less light at high angles than at low angles because of interference effects. This is caused by the fact that light scattered from one part of the molecule travels a different distance, and therefore is not exactly in phase with the light scattered from another part of the molecule. This phenomenon is quantified by defining the light scattering form-factor:

$P(\theta)$ = scattered intensity at angle θ / scattered intensity at angle $0°$

Calculations show that $P(\theta)$ can be written as a series:

$$P(\theta) = 1 - 1/3(q^2 Rg^2) + \ldots\ldots\ldots \qquad \text{Equation (12).}$$

where:
$q = 4\, n \sin(\theta/2)/\lambda o$
Rg = the radius of gyration of the molecules
n = the index of refraction of the fluid
λo = the wavelength of light in a vacuum.

Using measurements taken at scattering angles of $\theta = 15°$ and $90°$, equation (13) and (14) can be derived from equation (12).

$$P(\theta) = 1 - 26.3(Rg.n/\lambda o)^2 \quad (\text{for } \theta = 90°) \qquad \text{Equation (13).}$$

$$P(\theta) = 1 - 0.897(Rg.n/\lambda o)^2 \quad (\text{for } \theta = 15°) \qquad \text{Equation (14).}$$

It is now possible to solve $P(\theta)$ at either angle and derive the Rg value. The current calculations are carried out using the Debye function for Coils and all errors and a full explanation of the calculations can be found in the paper by Ford et. (9)al.

There is a Minimal Effect on Angular Dissymmety Polymers with Rg less than 10 nm (Wavelength 680 nm.)

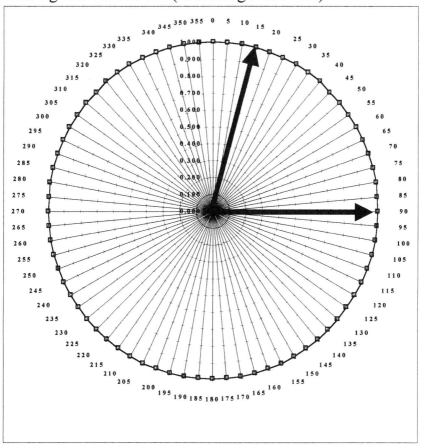

Figure 3.   The Debye function predicts that the particle scattering function for a particle less than 10 nanometers can be approximated to 1.

Branching Analysis of Polyolefins using Log Rg versus Log Mw
The same rules apply to light scattering as exists with the Mark Houwink relationship in viscometry.

$$Rg = K\,M^{\alpha}$$
Equation (15).

The exponent α is considered to signify linearity at between 0.57 and 0.61 for a random coil. How to measure this value is critical. In order to obtain a good Rg relationship to Mw, a number of key factors must be observed:

1. The signal to noise of each detector must be maximized. This is accomplished in a number of ways optically and through expert chromatography.

2. The optical approach established in the PD2040 produces the strongest signal possible by combining the technique of focusing the laser beam into a very small flow cell (10 µl) of optical volume (0.01 µl) and collecting the signal over a large solid angle. This produces a very high quality signal at both high and low angles as the scattering volume is minimized to approximately 0.01 µl and the cell design produces a plug flow at a flow rate of 1 ml per minute.

3. A stray particle traveling through scattering volume produces a large but very short signal. In the worst case, if the particle traveled only through the optical volume of the cell, the period that the particle would produce a high frequency signal would be for 0.5 seconds, but this rarely happens. The probability of a particle entering the beam is about 0.1%. Therefore, by collecting data at fast rates (100 pts/sec) and applying noise rejection algorithms, quality signals can be produced at averaged collection intervals of 1 point per second. Essentially, using either digital signal processing, or any 486 processor, this objective can be achieved.

4. The PD2040 cell design, using separate windows for collection in a matte black Teflon coated cell also reduces the amount of stray light entering the collection optics. This is critical, as the scattered light due to 1-2-4-trichlorobenzene has a Raleigh scatter of 35.7×10^{6} cm^{-1} and produces 29.75 times more scattered light than pure water and 8.5 times the scattered light than Tetrahydrofuran (THF). Therefore, the cell must be very efficient in collecting *only* the light produced from the optical volume as all other reflected light reduces the signal-to-noise ratio.

Any of the advantages afforded by optical design will be lost if good laboratory practice is not observed by maintaining a clean GPC/SEC system with particle-free columns.

High quality signals offer the ability to normalize the detectors to the NBS1482 polyethylene standards which has a molecular weight of 13,600 Daltons (Fig.4), providing excellent molecular weight analysis and the ability to measure Rg at relatively low molecular weights above 100,000 Daltons for the metallocene's polyolefins.
The calculation of the g-factor is not so easy. It is dependent on determining the Rg of the linear molecular weight of the polyolefin copolymer at the same molecular weight. Therefore, in (Fig.5) we compare the Rg versus molecular weight for a number of polymers: a linear polyolefin copolymer, the metallocene copolymers, and

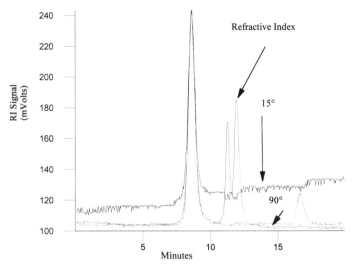

Figure 4. The use of the NBS standard 482 can be used to normalize the light scattering detectors as plenty of signal is available with an injection volume of 100 microliters of 3 milligram per milliliter solution.

Comparison Of Linear PE, PP, with Metallocene A and B

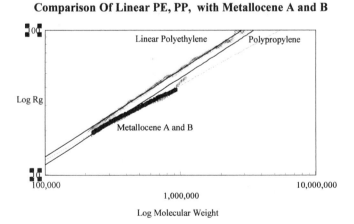

Figure 5. The metallocenes have very different slopes to that of the linear polyethylene and the polypropylene, indicating branching.

polypropylene. Without knowing the chemical composition of the metallocene's co-monomer ratio or whether the reaction is an alternating or random copolymer, it is difficult to determine what denominator to use for the g-factor calculation.

The Rg value is obtained from equation 13 and is often denoted as $<S^2>$
This is described as the root mean square radius of gyration. For calculation purposes Rg $= <S^2>$, Scholte has reviewed this extensively (12).

$$g = <S^2>_{br} / <S^2>_{ln} \qquad\qquad \textbf{Equation (16).}$$

For Rg$_{ln}$ value we assumed the value for polypropylene, as the composition of this polyolefin was known. This is not correct, but gives a good fit for g and demonstrates the fact that the metallocene is branched, although this particular technique only works above about 15 nm. in size (Fig. 6,7).

Results

The Viscometry results indicate that metallocene A is less branched than B (Table 1). By observing the *Mark-Houwink* exponent, the problem presented is that, although B is obviously branched, it is difficult to tell whether there really is any branching in A. The light scattering results demonstrate that there is no branching in the polypropylene or the linear polyolefin (Fig. 8, 9) but the exponent of 0.51 (Fig. 10) and 0.49 (Fig. 11) for Metallocine A and B (Fig. 12) respectively indicate a real difference from the other materials.

Conclusion

The viscometer was able to detect the effect of long chain branching in the different polymers. The issue that the metallocene polyolefins are relatively low in molecular weight contributed to the difficulty of this measurement. The use of the light scattering detector to identify long chain branching as a function of the different form factors produced by the effective size of the molecules (Rg) was much more apparent. This is very important when trying to detect branching in lower molecular weight polymers (less than 150,000 as there are, by default, less branching points available.)

The light scattering exponent was lower than expected for the metallocene polyolefins. The data conclusively demonstrates that, when compared with a linear polyolefin, polypropylene, and to the polystyrene standards, branching is very evident in these classes of metallocene's when using Rg calculated form light scattering measurments in place of intrinsic viscosity. The advantage of Rg is that it is very sensitive to branching at high molecular weight (Fig. 13.)

The advantage of viscometry is that it is still sensitive to the lower molecular weight region Branching is minimal and it is possible to estimate the exponent for the linear polymer, making the g[1] calculation possible without having to analyze an exact linear analogue of the branched polymer. This form of analysis will now be dependent on the user defining linear versus branched region for the viscosity measurement within the distribution.

Polypropylene

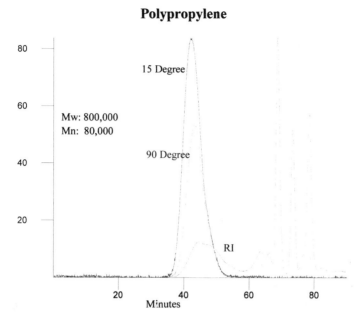

Figure 6. The polypropylene sample was measured using 20-micron packing material to minimize shear degradation from the columns.

Polypropylene

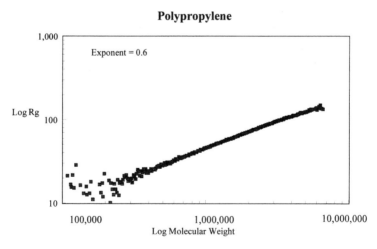

Figure 7. The exponent calculated for the Log Rg. versus Log Mw. plot indicates that there are no branches which is to be expected.

Table 1   A Comparison Between Branching Measurements Using GPC/Viscometry and GPC/Light Scattering

Polymer	Mw. Visc.	Mw. LS.	Visc. a	LS. a
Metallocene A	99,000	90,000	0.675	0.529
Metallocene B	100,000	98,000	0.63	0.489
Linear Polyethylene	135,000	138,000	0.7	0.602
Polypropylene	N/A	800,000	N/A	0.603

The comparison between the molecular weight averages from light scattering and viscometry are in very close agreement.
The exponents calculated from equations 4,15 give (a) values that indicate that both viscometry and light scattering when
combined with GPC can detect branching in the metallocenes.

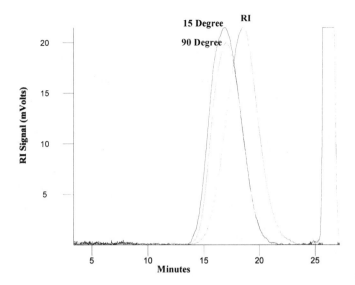

Figure 8. The linear polyethylene is run at low concentration to minimize viscose fingering. Injection volume of 100 microliters at a concentration of 1 milligram per milliliter.

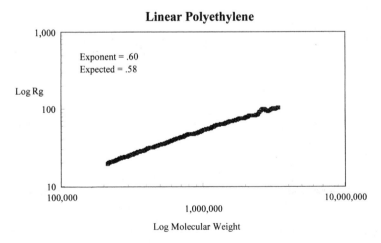

Figure 9. The exponent 0.6 calculated for linear low-density polyethylene indicates that there is not any branching in the heterogeneous catalyzed polymer.

246

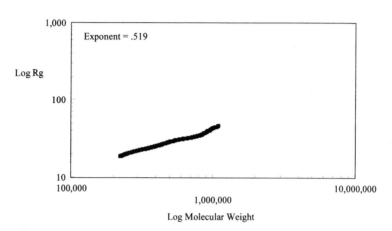

Figure 10. The exponent of 0.519 demonstrates that light scattering can detect branching in metallocene A.

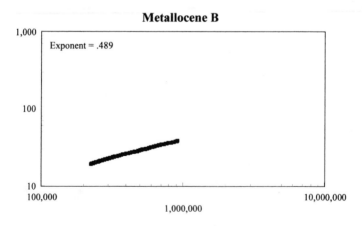

Figure 11. The exponent of 0.489 demonstrates that light scattering can detect branching in metallocene B.

Metallocene A and B

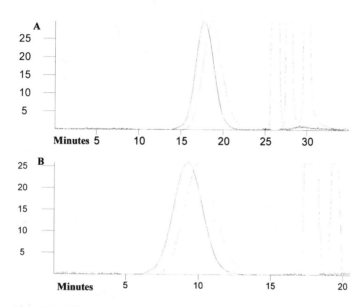

Figure 12. The difference in the particle scattering function can be observed for the normalized light scattering measurements. It is this difference that is used to calculate the Rg at each slice.

The g Factor Calculated for Metallocene A

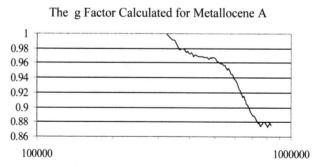

Figure 13. We used the data obtained from the polypropylene to demonstrate that a g factor can be calculated for branching. In order to get accurate results a metallocene catalyzed linear analogue is really required. The linear low density polyethylene did not contain enough octene co-monomer to give a similar hydrodynamic volume and therefore could not be substituted for linear metallocene in the g factor calculation.

There are a number of developments that aid in the investigation of long chain branching of these types of metallocene polyolefins at high temperatures; There has been a significant improvement in signal-to-noise from the new retrofit viscometers available, compared to the externally fitted Viscotek 100 series which was used for this work. Careful consideration will have to be given in plumbing these units in order to avoid inter-detector assumptions. Combining the molecular weight measurements from the low angle light scattering detector with the intrinsic viscosity from the viscometer detector will enable very accurate g^1 to be determined independently of the universal calibration which will eliminate another source of error. The main objective should always be to obtain the maximum sensitivity from all of the detectors. This is also a function of the condition of the chromatography system used.

References

1.   A. A. Montagna Designed Polyolefins p44 Chemtech Oct 1995 ACS Publication
2.   S.T. Balke, P. Cheung, R. Lew, T.H. Mourey. International GPC Symposium Proceedings, 1991. p. 117, Waters Publication.
3.   J. Lesec, M. Millequant. International GPC Symposium Proceedings, 1991. p. 221, Waters Publication.
4.   Zimm, B. H. and Stockmayer, W. H. J. Chem. Phys.,17 1301 (1949)
5.   A. E. Hamielic, H. Meyer. Online Molecular Weight and Long-Chain Branching Measurement using SEC and Low-Angle Laser Light Scattering. Chapter 3, Developments in Polymer Characterization – 5, 1986.
6.   Hert, M., Strazielle, C., Etude de la structure ramifiee du polyethylene basse densite par viscosite et difussion de la lumiere de fractions issues de la chromatographie d'exclusion GPC, Makromol. Chem., 184,135-145, 1983.
7.   N. Ford, T. Havard, P. Wallace Chapter 5, Particle Size Distribution III Assessment and Characterization. ISBN 0-8412-3561-9.
8.   B.H. Zimm. Journal Chem. Phys., 16, 1099 (1948).
9.   N.C. Ford, L. Frank, R. Frank. in Hyphenated Techniques in Polymer Characterization, Provder, Ed. ACS Advances in Chemistry Series, Vol. 247, American Chemical Society Series, 0065-23-93/95/0247-0109.
10.   T. Mourey and H. Coll. Journal Applied Polymer Sci., Vol. 56, 65-72, 1995.
11.   P. Debye. Journal Applies Phys., 15, 338 (1944).
12.   Scholte Th. G . Characterization of Long Chain Branching in Polymers Chapter 1 p3. - Developments in Polymer Characterisation – 4, 1984

Chapter 18

Characterization of Polyesters and Polyamides Through SEC and Light Scattering Using 1,1,1,3,3,3-Hexafluoro-2-propanol as Eluent

Antonio Moroni[1] and Trevor Havard[2]

[1]Boston Scientific Vascular, 45 Barbour Pond Drive, Wayne, NJ 09470
[2]Precision Detectors, Inc., 10 Forge Park, Franklin, MA 02038

ABSTRACT

The use of HFIP as eluent for the SEC characterization of traditionally difficult to dissolve polyesters and polyamides has several advantages over traditional solvents such as m-cresol and o-chlorophenol. HFIP readily dissolves these polymers at room temperature and does not require high temperature operation. HFIP is an aggressive organic solvent that behaves like a strong acid, therefore safety and instrument considerations must be kept into account. Issues such as safe operation procedures, solvent purification, column type, system calibration, detectors sensitivity must be dealt with before obtaining accurate and reliable results from the analysis. The most commonly used standards for system calibration are polymethylmethacrylate narrow molecular mass samples. However, the universal calibration concept does not always work in HFIP therefore, additional detecting techniques, such as Light Scattering and/or Viscometry, must be used to determine a polymer absolute molecular mass values. Light scattering offers the additional advantage of showing the conformation of a polymer molecule in solution and of allowing the measurement of its radius of gyration. This paper will describe how the authors were able to address the above considerations using an integrated SEC/Viscometry/Light scattering system based on the Waters 150CV GPC instrument.

INTRODUCTION

Size exclusion Chromatography of engineering thermoplastics of the family of polyesters and polyamides can only be performed in a limited number of solvents, each of which has characteristic advantages and disadvantages, as shown in below.

SOLVENT	REMARKS
Orthochloro Phenol or Methacresol	Inexpensive but dangerous solvents. Because of their high viscosity, analysis must be performed at high temperature, where polymer degradation may occur. This system requires heating between 140° and 160° C.
Methylene Chloride - Dichloroacetic Acid	Inexpensive but highly corrosive. Some samples must be dissolved at high temperature, where polymer degradation may occur.
Chloroform - HFIP	Somewhat expensive, dangerous, chloroform suspected carcinogen. It appears to be corrosive to metals used in instrumentation tubing. This mixture does not dissolves some polyesters or polyamides. It allows the use of polystyrene standards.
Methylene Chloride - HFIP	Fairly expensive. Forms azeotrope boiling at 30 C. May cause formation of vapor lock in the instrument. This mixture does not dissolves some polyesters or polyamides. It allows the use of polystyrene standards.
1,1,1,3,3,3-Hexafluoro-2-Propanol (HFIP)	Very expensive, dangerous, highly polar solvent. Dissolves several polymers easily at room temperature. A ion pairing agent may be necessary to prevent non-size exclusion interaction of certain polymers with some column packing. Polystryrene standards cannot be used. Polymethyl methacrylate standards can be used but analysis results may be affected by the presence of ion pairing agent or water in the solvent.

HFIP appears to be the solvent most convenient to use, although it is expensive and somewhat irritant. The drawback of this system is the difficulty to obtain absolute molecular mass values because Universal Calibration has been reported not to work with this system and polystyrene standards are insoluble.

Therefore, the objective of this paper is to evaluate the performances of a chromatographic system dedicated to polyester and polyamides analysis using HFIP as the effluent , to test the validity of the universal calibration concept, and to propose an alternative method to determine the absolute molecular weight of polymers soluble in HFIP.

BACKGROUND

The Universal Calibration concept is based on the assumption that SEC separates macromolecules by size and that different polymers may be all placed on the same curve if a measure of molecular volume is used, rather than molecular mass (1,2,3). For Gaussian coil polymers, the molecular volume can be expressed in terms of the polymer intrinsic viscosity $[\eta]$ times the molecular mass of the polymer through the Fox-Flory relation
(4):

$$[\eta] = \Phi <r^2>^{3/2} / M \qquad [1]$$

where Φ is the universal Flory viscosity constant and $<r^2>$ the mean square end-to-end distance.

Molecular Mass determination of polymer distributions can be calculated according to the Universal Calibration theory using on-line viscometry in conjunction with conventional SEC, to measure both the specific viscosity and the mass concentration of each fraction of solvated polymer, as it elutes from the separation columns. At the very low concentrations used in SEC, the specific viscosity of the polymer solution as it elutes can be expressed as the product of its concentration times its intrinsic viscosity:

$$\eta_{sp} = c_i \, [\eta]_i \qquad [2]$$

Intrinsic viscosity of a polymer can be related to its average molecular mass through the well known Mark-Houwink-Sakurada equation :

$$[\eta] = K \, M^a \qquad [3]$$

Light scattering measurement is another way to determine the absolute molecular mass of polymers (5,6). Furthermore, by measuring the amount of light scattered at at least two angles, it is possible to calculate the mean square radius of gyration ($<r_g^2>$) and the radius of gyration ($r_g = <r_g^2>^{1/2}$) of the polymer molecules, quantities that describes the average distance of the molecule component from the center of mass(7).

Light scattering from a polymer molecule (7) is described by the Raleigh equation :

[4]
$$\frac{K^* c}{R (\Theta)} = (1/Mw)[(1+16\pi^2/3\lambda_0^2 \ ) <r_g^2> \sin^2 (\Theta/2)] + 2 A_2 c$$

where

$$K^* = 4\pi^2 (dn/dc)^2 n_0^2 / (N_A \lambda_0^4) \qquad [5]$$

and c is the concentration of the solute molecules (g/mL), R (Θ) is the fraction of light scattered at angle Θ relative to the intensity of the incident beam, N_A is Avogadro's number, λ_0 is the wavelenght of the light, n_0 the refractive index of the solvent, dn/dc is the refractive index increment, which tells how much the refractive index of the solution varies with solute concentration. Mw is the weight-average molecular mass and A_2 is the second virial coefficient, a measure of solvent-solute interactions that is not important in SEC application because of the low concentrations used. A solvated macromolecule conformation can be measured by relating its Mw with rg over a range of molecular mass. SEC/LS, separating molecules by size, and measuring light scattering and concentrations of each slice, offers an ideal way to accomplish the above.

Molecular mass for a polymer can be calculated across all increments of a chromathogram using the Debye plot of R (Θ)/Kc versus $\sin^2 (\Theta/2)$. The intercept of the curve is Mw while the slope is the mean square radius of gyration $<rg^2>$.

Rg varies with Mw according to the following equation:

$$rg_i^\alpha \propto Mw_i \qquad [6]$$

where α = .33 for spheres, .5 for random coils at the Theta point, and 1 for rigid rods.

Therefore, the slope of a plot of LOG (rg) vs. LOG (Mw) indicates the conformation of the solvated molecule.

EXPERIMENTAL

Analysis were run in a Waters 150 CV chromatograph, equipped with a Waters differential refractometer, in line single capillary viscometer, and a two angles light scattering.

The self contained 150 CV is an ideal instrument to analyze difficult polymers under difficult conditions, such as high temperature, and with dangerous solvents. The large column chamber of the 150 CV can house several columns and detectors under thermostatic conditions and in a sealed environment. The 150 CV is equipped with on-line single capillary viscometer and allows continuos monitoring of the specific viscosity of the effluent. The effluent the is carried to the on-line differential refractometer to determine polymer concentration. The inter-detector volume was measured by running a series of narrow molecular mass PMMA standards and measuring the time delay between the two detectors response peaks. For our particular instrument, it was found to be 140 μL.

The light scattering detector used is the Precision Detector model 2020. It can measure light scattering at angles of 15 and 90 degrees , has a very small detection chamber, thus minimizing chromatographic band broadening, and fits precisely inside the 150 CV thermostated chamber. To minimize dead volume, it was inserted between the columns and the viscosity detector.

Several types of columns were alternatively used to perform the separation, including Shodex model HFIP 803 and 804, Jordi mixed bed gel, and Polymer Laboratories Mixed C and HFIP Monodisperse columns. The best overall results were obtained with either the Jordigel or the P.L. Monodisperse column. The columns were kept at 40° C constant temperature in the 150 CV column chamber.

Samples to be analyzed were prepared at a concentration of 1 mg/ml in HFIP, either with or without Sodium trifluoro acetate as a ion pairing agent, in concentration of .01 M or .02 M. The addition of this limited amount of ion pairing agent was sufficient, in most cases, to avoid non-size exclusion interaction of the polymer molecules with the column packing, and to maintain the polymer peak shape undistorted. Addition of the ion pairing agent to the whole amount of solvent running through the instrument was avoided, to prevent valve blockages and general clogging of the instrument.

Narrow molecular mass samples concentration was an inverse function of molecular mass and was adjusted to keep the viscometer signal between 10 and 20 mV, to remain in the linear range of the instrument.

Broad samples were prepared at a carefully measured concentration of about 1 mg/mL. In the case of polyester samples, the trailing peak, attributed to cyclic oligomers impurities, was excluded from the calculation to determine the moments of the distribution.

Analysis were run at flow rates of between .6 and 1.2 ml/min, depending on the column used. A flow rate setting of .8 ml/min (actually .74 ml/min) was found to provide the highest resolution. Flow constancy was insured by monitoring the retention time of the water peak, always present in HFIP as an impurity.

Polymer standards were obtained from Polymer Laboratories Inc., 160 Old Farm Road, Amherst, MA 01002 and from American Polymer Standards Corporation, 8680 Tyler Blvd., Mentor, OH 44060.

The HFIP carrier solvent was purchased from Aldrich, Hoechst-Celanese, and Nu Brand in one gallon lots. It was dried and purified by distillation (b.p. 59° C) over Barium Oxide, then filtered through .5 μm PTFE filters. Because of its high cost, HFIP was recycled through the instrument and to a sealed container filled with 500 - 1000 ml of it. The solvent was frequently redistilled to prevent accumulation of impurities. Extreme care was taken to protect the operator from contact with HFIP. In fact, this solvent is irritant to mucous membranes, caustic to the skin, and extremely destructive to the corneal membrane.

RESULTS AND DISCUSSION

The "Universal Calibration" curves obtained from several polymers dissolved in HFIP do not overlap, as shown in Fig. 1. The curves, obtained on the PL HFIP Monodisperse column, appear shifted and with different slopes. PMMA narrow molecular mass standards curve is linear between 3800 and 496,000, showing exclusion above and a deviation from linearity below these values. The addition of a ion pairing agent to the PMMA samples reduced the exclusion limit to 333,000, but had no other significant effect on the overall shape of the curve.

Viscosity Law curves show slopes of about .5 for molecular masses of less that 10,000 while slopes for higher molecular masses range from .6 to . 9, depending on the polymer, as shown in Fig. 2 and reported in Table 1. The value for **a** of .5 for low molecular mass polymers has been observed for polymers of average molecular mass lower than 10,000 (9). For high molecular mass polymer, the values for the **a** coefficient of the Mark Houwink equation suggest that these polymers are arranged in a number of conformation in solution, ranging from a fairly compact coil for the flexible Polyamide 6 (**a**=.6) to a fairly extended rod for Poly(oxyethylene) (**a**= .92). This last polymers also show extremely high viscosity values in HFIP. Poly(oxyethylene) chains have already been reported to be quite extended in a hydrogen bonding solvent such as water (**a** = .82) (12) .

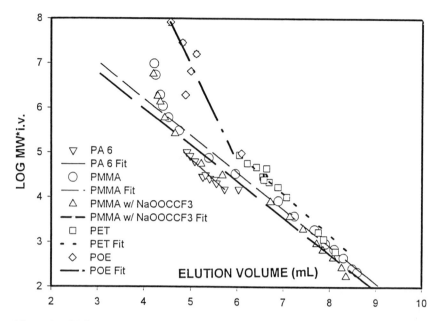

Figure 1.   Universal Calibration Curves of Polymers in HFIP. P.L. HFIP Monodisperse Column.

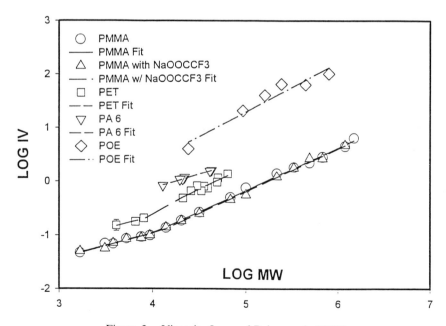

Figure 2.   Viscosity Laws of Polymers in HFIP.

TABLE 1. Universal Calibration and Viscosity Data for several polymer in HFIP.

Polymer	U.C. Constant	U.C. Slope	Viscosity Law $K*10^3$ (mL/g)	Viscosity Law a	Notes
Polyamide 6 (withIon Pairing Agent, .01M)	10.18±.13	-1.06 ± .14	4.37	.59 ± .1	
PMMA ($<10^4$) ($>10^4$)	11.52 ±.13	-1.21 ± .03	1.55 .07	.46 ±.04 .79 ±.02	U.C. fitted between 3800 and 496000 MM
PMMA (with Ion Pairing Agent, .01M) ($< 10^4$) ($> 10^4$)	11.36 ±.16	-1.23 ±.04	1.38 .06	.48 ±.05 .80 ± .02	U.C. fitted between 3800 and 333000 MM
PET ($< 10^4$) ($> 10^4$)	11.45 ±.05	-1.09 ±.06	4.17 .13	.43 ±.09 .83 ±.04	U.C. fitted for MM > 10,000
POE	8.19 ±.26	-.45 ±.11	.51	.92 ±.12	

Poly (methyl methacrylate) (**a**= .79) and Poly (ethylene terephthalate) (**a**= .83) chains appear also quite extended in HFIP. These two results are also confirmed by Light Scattering analysis, where the slope of LOG(r_g) vs. LOG (MW) shows values of α = .591 for PMMA both narrow and broad MM standards (in the range of MM > 200,000, Fig. 3), and α = to .63 for PET (in the range above 30,000, averaged from at least four measurements for each of three PET standards with Mw of 39,000, 49,200, and 63,500). Results obtained from a PET 63.5 K standard are shown in Figures 4,5 and 6.

The viscosity **a** value can be related to α values from light scattering data through the equation (7), derived from the Ptitsyn-Eizener (10) and Mark-Houwink-Sakurada equations (11):

$$\alpha = (a + 1) / 3. \qquad (7)$$

Table 2. shows the experimental values of **a** and α, and the α value calculated accordingly to Eq. (7) for PET and PMMA. The similarity between α_{exp} and α_{calc} is remarkable. Thus, these two techniques independently confirm the extended conformation of the polymer molecules solvated in HFIP

Determination of average molecular mass values determined for PET samples through a Universal Calibration curve derived from narrow PMMA standards gives grossly underestimated results, as shown in Table 3 and 4. Results obtained using another Universal Calibration curve based on broad PET standards gives better results, although using broad distribution standards to calculate the parameters of the $[\eta]*MW$ relation versus elution volume can give unreliable results because of the diffusion that the distant part of the distribution may have undergone during SEC separation. This effect tend to increase the apparent polydispersity and lower the value of the Mark-Houwink exponent **a**. In fact, this may be the explanation for the lower value of **a** found when the viscosity law is fitted within a PET sample distribution, as detailed in Figure 7. Therefore, more than one broad standard should be used to build a meaningful calibration curve. This effect is even more pronounced with PA 6, which tends to have considerable non-size exclusion interaction with the column packing, and to give a molecular mass distribution curve skewed toward the low end. On the other hand, average molecular mass values obtained by Light Scattering tend to give values in agreement with manufacturer's specifications. In fact PET relatively high dn/dc in HFIP (.235) affords a high signal to noise ratio and allows light scattering measurements to reliably extend to the low end of the molecular mass distribution, thus allowing a meaningful calculation of Mn.

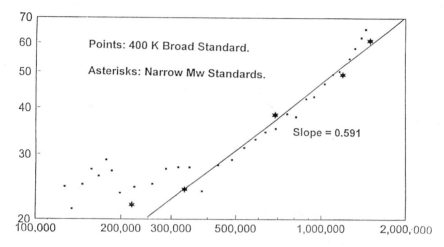

Figure 3. PMMA in HFIP: Log Rg versus Log Mw Curve. 0.01 M NaOOCCF₃ in Sample.

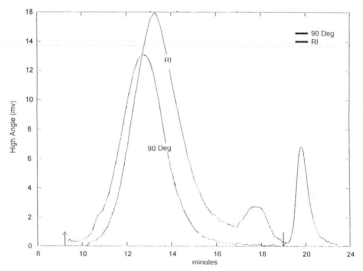

Figure 4. Refractive Index and 90° Light Scattering Output for PET 63.5 Mw Standard.

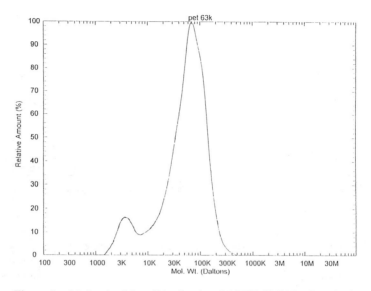

Figure 5.   Molecular Mass Distribution Of PET 63.5 Mw Standard.

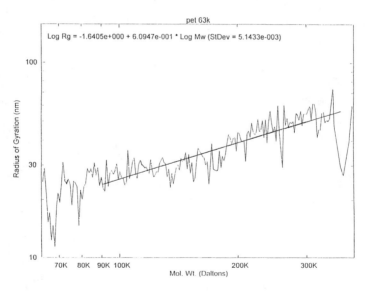

Figure 6.   Radius of Gyration as a Function of Molecular Mass for PET 63.5 Mw Standard.

TABLE 2. Values of exponents **a** and α for polymers in HFIP.

Polymer	**a** (exp)	α (exp)	α (calc)
PMMA	.79	.591	.597
PET	.83	.634	.610

TABLE 3. Average Molecular Mass Values for PET 39 K Standard.

PET 39 K	Manufacture's Specifications	Light Scattering	Universal Calibration (PMMA Standards)	Universal Calibration (PET Standards)
Mn	21,000	18,500	7,999	17,253
Mw	39,000	36,900	11,520	40,300
Mz	69,000	52,000	16,230	64,880

TABLE 4. Average Molecular Mass Values for PET 49K Standard.

PET 49 K	Manufacture's Specifications	Light Scattering	Universal Calibration (PMMA Standards)	Universal Calibration (PET Standards)
Mn	25,600	23,700	8,060	22,660
Mw	49,200	55,000	11,500	50,700
Mz	72,800	72,000	16,150	70,350

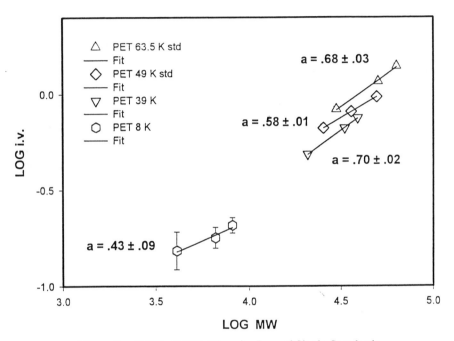

Figure 7.   PET in HFIP: Viscosity Law of Single Standards.

CONCLUSIONS

The Universal Calibration concept does not work well for polymers solvated in HFIP and the different solvation states and chain conformations that different polymers show may be the reason. Non-size exclusion interactions between macromolecules and column packing may be an additional reason for this behavior. Macromolecules rich in polar groups become quite extended in HFIP, suggesting that they are solvated through hydrogen bonding. The hypothesized attachment of HFIP molecules to the polymers main or side chain may make the polymer stiffer, bulkier and more extended than normal, thus increasing its viscosity.

If the Universal Calibration concept is not applied, on line viscosity measurements do not help to determine the absolute molecular mass of different class of polymers, although it may give information on the solvated molecule conformation.

Light Scattering determines absolute molecular mass slice by slice during polymer elution, independently from flow rate variation, polymer-column packing interaction, and it still can give accurate results even in system where macromolecules elute independently of their molecular size. Molecular mass values obtained by light scattering for PET show the best agreement with manufacturer's supplied values. Light Scattering also allows r_g measurement, and r_g vs. Mw dependence also allows the determination of the macromolecule conformation in solution. Results obtained for PMMA, both narrow and broad molecular mass standards, and for PET, agree with viscosity measurements and confirm that both molecules are quite extended in HFIP.

REFERENCES

1) Grubisic, Z.; Rempp, P.; Benoit, H.; J. Polym. Sci. , Polym. Lett. Ed., B5, 753-759 (1967).
2) Cardenas, J.N.; O'Driscoll, K.F.; J. Polym. Sci., Polym Lett. Ed., 13, 657-662 (1975).
3) Swartz, T.D.; Bly, D.D.; Edwards, A.S.; J. Appl. Polym. Sci., 16, 3353-3360 (1972).
4) Fox, T.G.; Flory, P.J.; J. Am. Chem. Soc., 70, 2384 (1948).
5) Tanford C.; Physical Chemistry of Macromolecules, John Wiley & Sons, Inc. New York (1961), Chapter 5.
6) Lechacaux, D.; Panaras, R.; Brigand, G. and Martin G.; Carbohydrate Pol., 5 423-440 (1985).
7) Kratochvil, P.; Classical Light Scattering from Polymer Solutions, 1st Edition, Elsevier, Amsterdam, 1987.
8) Flory, P. Principles of Polymer Chemistry, 1st Edition, Cornell University Press, Ithaca, New York, 1981.
9) Flory, P.J., Fox, T.J.; J. Am. Chem. Soc., 73, 1909 (1951).
10) Eizner, Y.E., Ptitsyn, O.B.; Vysokomol. Soedin., 4,1725,1962.
11) Wyatt, P.J., Anal. Chim. Acta; 272, 1 (1993).
12) Baley, F.E., Callard, R.W.; J. Appl. Polym. Sci. 1, 56, (1959).

Chapter 19

Degradation of Substituted Polyacetylenes and Effect of This Process on SEC Analysis of These Polymers

Jiří Vohlídal and Jan Sedláček

Department of Physical and Macromolecular Chemistry, Laboratory of Specialty Polymers, Faculty of Science, Charles University, Albertov 2030, CZ-128 40 Praha 2, Czech Republic

Survey of advances in the fields of: (i) spontaneous autoxidative degradation of substituted polyacetylenes, and (ii) effect of this process on SEC analyses of these polymers, involving already published and new original results is presented. The degradation is shown to be, essentially, of the random type but accompanied by the enhanced low-MW species elimination in case of high-cis polymers. These species are suggested to be formed in the reactive relaxation of excited ends of primary fragments resulting from the random oxidative cleavage of macromolecules. The degradation proceeds even in SEC columns and distorts results of the polymer SEC analysis. Theoretical treatment and computer simulation of such analyses provided information on the results' distortion extent and showed that systematically biased MW averages provide plausible degradation rate constants.

Size partitioning of macromolecules in SEC columns is more or less perturbed by various hydrodynamic and thermodynamic effects that do not alter a constitution of analyzed macromolecules (1-3). Only in case of very long macromolecules, typically with MW above $5 \cdot 10^6$, the degradation induced by hydrodynamic shear forces can occur in SEC columns (4,5). This process has been reported for various high-MW polymers like poly(styrene), polyethylene, poly(methyl methacrylate), poly(isoprene), poly(oxyethylene) and poly(isobutylene) (4-15) so that it is well established and known, nowadays. On the other hand, little is yet known on the effect of chemically induced degradation in the polymer SEC analysis that was recently demonstrated in the case of poly(phenylacetylene) (16) which undergoes rather fast oxidative degradation in solu-

Supported by the Ministry of Education of the Czech Republic, Project VS 97103.

tions. This phenomenon together with the oxidative degradation of selected ring-substituted poly(phenylacetylene)s are discussed in the present paper.

SEC Study of Degradation of Substituted Polyacetylenes

Photonic and electrical properties of substituted polyacetylenes are mostly investigated on thin films or sandwiched layers prepared by controlled casting from polymer solutions *(17-23)*. If an investigated polymer is not resistant to oxygen, it can undergo oxidative degradation deteriorating its functional properties when exposed to air during the film preparation and/or physical measurements. Therefore, knowledge of stability or degradability in air of a given polymer in both the solid state and solution is crucial for optimization of the casting technique and physical measurement conditions and, also for an adequate interpretation of results of physical measurements.

Effectiveness of SEC Method in Investigation of Polymer Degradation. Polymer autoxidative degradation can be detected and monitored by various methods differing in the type of yielded information and detection limit. *Oxygen uptake measurements* provide indirect information about the polymer oxidation including kinetics of this process (see, e.g., autoxidation of poly(methylacetylene) *(24)*). *Vibrational and NMR spectra* provide direct qualitative and quantitative evidence of polymer oxidation (occurrence and intensity increase of bands and signals of oxygen containing groups) *(24 - 26)*. *ESCA and related methods* provide information on chemical composition of polymer surface layers *(25)*. *UV spectra* provide information about length of conjugated sequences (Figure 1) that is, however, not in a direct relation to the main chain length *(24, 26)*. *ESR spectra* are helpful mainly in revealing the autoxidation mechanism *(24, 27)*. It should be pointed out, that all the above listed methods can provide only indirect evidences of polymer degradation because none of them can detect a decrease in the polymer MW value.

Unlike the preceding methods, *SEC technique* can provide a direct evidence of the polymer degradation - the information on a drop in the polymer MW value. Moreover, the SEC method is significantly more effective (sensitive) in detecting a degradation than the above indirect methods. Let us, for example, consider the decrease in the polymer number-average degree of polymerization from the initial value $<X>_n^0$ = 4 000 to final value $<X>_n$ = 2 000, which is due to oxidative degradation yielding fragments with easy detectable carbonyl end groups. There is no doubt that SEC will easily and reliably detect the 50 % drop in MW. However, the mole fraction of carbonylated monomeric units in the final polymer will be equal to 0.0005, far below the detection limit of spectroscopic methods. Reliable detection of carbonyls requires ca ten times higher concentration, which corresponds to a drop in $<X>_n$ to ca 1/10 of the original value (see, e.g., results in ref. *(26)*). Advantage of SEC method in detecting the early stage of polymer degradation is thus obvious.

SEC technique still offers other useful information about the polymer degradation. It visualizes the polymer MW distribution changes due to degradation, which is important for disclosing the position mode of macromolecule scission (random, midpoint, chain-end or other mode) *(26, 28-36)*. In addition, SEC makes available the time dependences of MW averages that are needed for the kinetic quantification of degradation process.

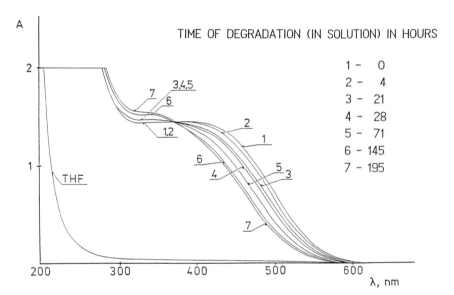

Figure 1. UV/vis spectra of poly(phenylacetylene) sample (PPA/W) degrading in THF solution at room temperature monitored at indicated degradation times. Adapted from ref. *(26)*, Copyright 1993, with kind permission of the Institute of Organic Chemistry and Biochemistry, Academy of Sciences of the Czech Republic, 166 10 Prague 6, Czech Republic.

Kinetics of Polymer Degradation. This kinetics is generally described by the Simha-Montroll equation *(33, 34)*, that express how the number of polymer chains, N_x, with a degree of polymerization, (DP), equal to X varies with the degradation time, t:

$$\frac{dN_x}{dt} = -\sum_{p=1}^{X-1} k(X,p)N_x + 2\sum_{S=X+1}^{\infty} k(S,X)N_s \tag{1}$$

where the rate constants $k(X,p)$ and $k(S,X)$ give the probability of cleaving the chain of length X (or S, first variable) at a distance p (or X, second variable) from the chain end.

When dealing with the chemically induced polymer degradation, the random and chain-end modes of a main-chain cleavage come mainly into consideration. Acidic hydrolysis of polysaccharides is a typical example of the random polymer degradation because the choice of a main-chain link that will be cleaved in the next reaction step is of the random nature. Depolymerization of macromolecules with active ends taking place in the systems tempered above the polymer ceiling temperature is typical example of the chain-end mode degradation.

Random Cleavage. If each of $(X-1)$ main-chain links of any macromolecule in the system is equally accessible to scission, then $k(X,p) = k(S,X) = v$ and the Simha-Montroll equation can be written as:

$$\frac{dN_x}{dt} = -(X-1)v\,N_x + 2v\sum_{S=X+1}^{\infty} N_s \tag{2}$$

where v is the rate constant of a main chain bond cleavage that is, the probability that a randomly selected bond will be cleaved within a unit time interval. The first term in equation 2 represents the rate of decay of X-mer macromolecules and the second term the rate of their simultaneous formation in the cleavage of longer macromolecules ($S > X$).

Analytical solution to equation 2 yields the following time dependences of the DP averages: $<X>_n^t$, $<X>_w^t$ and $<X>_z^t$, of a randomly degrading polymer *(34,36)*:

$$\frac{1}{<X>_n^t} = \frac{1}{<X>_n^0} + vt \tag{3}$$

$$\frac{1}{<X>_w^t} = \frac{1}{<X>_w^0} + \frac{v}{3}\int_0^t I_w(t)dt \tag{4}$$

$$\frac{1}{<X>_z^t} = \frac{1}{<X>_z^0} + \frac{v}{2}\int_0^t [I_z(t) - \frac{2}{3}]dt \tag{5}$$

Here $I_w(t) = <X>_z^t/<X>_w^t$ and $I_z(t) = <X>_{z+1}^t/<X>_z^t$ is the polydispersity index based on the weight-distribution and z-distribution of DP, respectively. The integrals on the right hand sides of equations 4 and 5 are obtained by numerical integration of measured I_w vs t and I_z vs t dependences. In the case of purely random degradation, all three time dependences given by equations 3 to 5: (i) should be linear; and (ii) should yield the same value of the degradation rate constant v. If one of these criteria is not fulfilled, the polymer degradation cannot be regarded as to be of the purely random type and another mode of the main-chain cleavage (midpoint, chain-end or a more complex mode *(31)*) and/or the fragments cross-linking should be taken into consideration.

Chain-End Cleavage. Consider the simplest example of the chain-end cleavage mode, the depolymerization of a high-MW polymer free of oligomers running without a recombination of fragments and volatilization of nondegradable, low-MW products. In that case the total number of all high-MW X-mers is remaining constant being equal to the initial number of macromolecules in the system, N_0. However, the overall number of molecules in the system, N, increases due to formation of low-MW fragments in the course of depolymerization. Assuming a uniform rate constant of chain-end scission, v_e, for all high-MW X-mers, the increase in N is described by the rate equation: $dN/dt = v_e \cdot N_0$, which means that the total number of molecules in the system linearly increases with the degradation time: $N = N_0(1 + v_e t)$. Consequently, the reciprocal value of $<X>_n$ will rise linearly with degradation time:

$$\frac{1}{<X>_n^t} = \frac{1}{<X>_n^0}(1 + v_e t) \qquad (6)$$

Examples of Experimental Observations. First detailed investigations of the autoxidative degradation of substituted polyacetylenes were performed with poly(methylacetylene) *(24)* and poly(phenylacetylene), PPA, *(25)*. Carbonyl and hydroxyl groups were detected as the main products of these autoxidations. In these studies, however, main attention has been paid to the chemistry of this degradation and the SEC method was used as a supporting technique only. Later studies on PPA *(26, 37)* as well as our recent studies on PPA derivatives (see later) that were mainly focussed at the kinetics of this process have been performed by using SEC as the main experimental method. The reported measurements were performed using a Tsp (Florida) chromatograph equipped with RI and UV (254 nm) detectors and a series of two columns: PL Mixbed-B and Mixbed-C (Polymer Laboratories, Bristol). The following conditions were applied: stabilized THF, flow rate 0.7 mL/min, injected 10 µL of polymer solution (conc. from 0.1 to 1 mg/mL), evaluation by the PS calibration curve method. First analysis of a sample was performed 20 min after mixing the polymer with THF. Consecutive analyses were made from the same solution stored in the meantime in air at room temperature. Examples of results of such studies are shown in next paragraphs.

Structure and Main Features of Degradation of Polyacetylenes. Substituted poly(phenylacetylene)s can be prepared in various structure forms differing mainly in the configuration of repeating units (Figure 2). In this paper, we will discuss two types of polymers: (i) those prepared WOCl$_4$-based catalysts, further denoted by the extension W, which have the high-trans structure *(38 -43)*, and (ii) those prepared

268

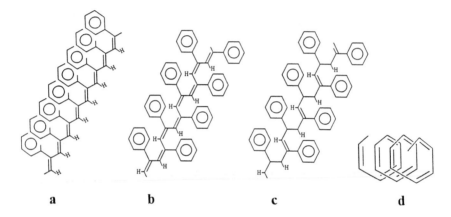

a   b   c   d

Figure 2. Planar projections of main configurational isomers of substituted poly(p-henylacetylene)s with a regular head-to-tail linkage of repeating units: a) trans-transoid, b) cis-transoid, c) trans-cisoid and scheme of the cis-cisoid polyacetylenic chain: d).

with Rh(diene) catalysts (extension /Rh), which are regarded as the regular, head-to-tail high-cis polymers *(27, 42 – 48)*. The W-type polymers are mostly dark red whereas those of the Rh type are yellow, which points to a higher extent of π-conjugation for the W-type polymers compared to those of the Rh-type.

Regardless of the actual polymer structure, all investigated monosubstituted poly(phenylacetylene)s were found to be stable in the year scale when stored in a sealed evacuated vial or under an oxygen-free inert atmosphere (see, e.g., Ref.*(26)*). However, most of them undergo degradation when exposed to air. Unlike the vinylic and other saturated polymers, substituted polyacetylenes degrade without any induction period, which indicates that oxygen is directly involved in the initiation processes *(16, 24-26, 37, 41)*. The degradation is generally very slow in the solid state but quite fast in aerated solutions, the difference in the rate being of the two orders in magnitude *(26)*.

$$\left[\begin{matrix} C = CH \\ | \\ \bigcirc \end{matrix} \right]_n$$

Poly(phenylacetylene) (PPA).   Time dependence of the SEC trace for high-trans PPA/W degrading in aerated THF solution (conc. 0.5 mg/mL) is shown in Figure 3. As it can be seen, the MW distribution of PPA/W sample remains unimodal during entire process. All three kinetic plots according to equations 3 to 5 are linear and yield almost identical values of the degradation rate constant: $v = (2.5 \pm 0.2) \cdot 10^{-6}$ min^{-1} for the degradation in non-stabilized THF *(26)* and $v = (2.0 \pm 0.2) \cdot 10^{-6}$ min^{-1} for the degradation in THF stabilized by 2,6-ditert.butyl-4-methylphenol (0.025 weight %) *(16)*. In summary, the autoxidative degradation of PPA/W suits the laws of the random polymer degradations.

Unlike the preceding case, the SEC trace of PPA/Rh sample changes from unimodal to trimodal showing an enhanced formation of two kinds of oligomeric species during the degradation (Figure 4). According to elution volumes and GC-MS analysis, these species correspond to monomer and cyclotrimers (triphenylbenzenes), respectively. Corresponding kinetic plots for this sample provide v values significantly depending on the type of used DP averages: $v_n = (4.5 \pm 1) \cdot 10^{-6}$ min^{-1} >> $v_w = (0.95 \pm 0.15) \cdot 10^{-6}$ min^{-1} > $v_z = (0.50 \pm 0.15) \cdot 10^{-6}$ min^{-1} (subscript indicates the type of DP averages used for a determination of the particular v value).

Both the increased formation of oligomeric species and non-uniformity of v values clearly demonstrates that the degradation of the high-cis PPA/Rh sample is not of the purely random type. As PPA/W and PPA/Rh differ in the microstructure only, there is not obvious reason for a principal difference of the mechanism of oxidative cleavage of their macromolecules. Therefore, the obtained results suggest that the degradation of PPA/Rh proceeds as the random cleavage of its macromolecules into fragments that partly undergo a consecutive elimination of monomer-type species and cyclotrimers, i.e., the chain-end cleavage. The rate constant of random

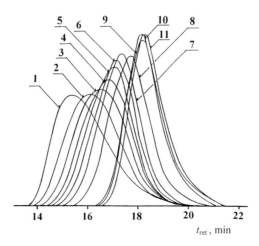

Figure 3. Time development of SEC records of PPA/W sample (prepared with WOCl₄/Ph₄Sn catalyst in benzene/dioxane, 1:2 by vol.) degrading in THF at room temperature; curve − degradation time: **1** − 35; **2** − 102; **3** − 162; **4** − 230; **5** − 300; **6** − 370; **7** − 430; **8** − 500; **9** − 1 510; **10** − 1 580; **11** − 1 650 min.

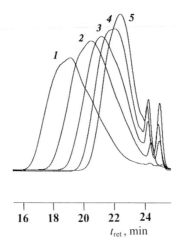

Figure 4. Time development of SEC records of PPA/Rh sample (sample prepared by Rh(*nbd*)(−C≡CPh)(PPh₃)₂ catalyst in THF) degrading in THF at room temperature; curve − degradation time: **1** − 37; **2** − 3 020; **3** − 7 530; **4** − 12 930; **5** − 23 000 min.

cleavage can be good approximated by the value of v_z ($0.5 \cdot 10^{-6}$ min^{-1}) because experimental $<X>_z$ values are virtually unaffected by a presence of low-MW fragments. On the other hand, the value of v_n should involve contributions of both the random cleavage rate constant (i.e., v_z) and the terms $v_e/<X>_{nf}^0$ (see Equation 6) characterizing formation of cyclotrimers and monomer-type species. It seems, however, impossible to estimate the values of v_e because the value of $<X>_{nf}^0$ refers to energetically excited primary fragments formed in the random scission only and not to the original polymer and already deactivated fragments (molecules).

$$\left[\!\!\left[\begin{array}{c} -C = CH \\ | \\ \bigcirc \end{array} \right]_{4.3} \!\!\left[\begin{array}{c} C = CH \\ | \\ \bigcirc \\ NO_2 \end{array} \right]_{1} \right]_n$$

Statistical Copolymer of Phenylacetylene (PA) and 4-Nitro-PA (NPA). The copolymer prepared with WOCl$_4$-based catalyst *(42)* and containing 4.3 PA units per one NPA unit exhibits another type of degradation behavior. The SEC records given in Figure 5 show only marginal formation of low-MW species and, in principal, their time development resembles that typical of the random degradation. However, kinetic dependences (Figure 6) show significant nonlinearity on the whole measured time interval indicating complexity of this degradation.

Degradation of this copolymer should be analyzed in terms of a variety of main-chain links in its macromolecules. Three types of links must be considered: PA-PA, PA-NPA and NPA-NPA that can differ in the reactivity to oxygen. The observed concave course of kinetic plots can be then explained as a result of the random degradation taking place simultaneously on both weaker and stronger links *(28)*. Accordingly, the initial rate of degradation is controlled by the cleavage of week links (most probably PA-PA), whereas the final rate (at high values of *t*) by the cleavage of stronger links (most probably PA-NPA because occurrence of NPA-NPA links should be rather rare). The estimated initial value of $v = 1.5 \cdot 10^{-6}$ min^{-1} is significantly lower than that found for PPA/W (see above) and, moreover, the value of v for high degradation times is at least about one order of magnitude lower. This indicates that the ring-substitution by nitro groups makes PPA more resistant to the oxidative degradation.

$$\left[\begin{array}{c} C = CH \\ | \\ \bigcirc\!\!\diagdown \end{array} \right]_n$$

I

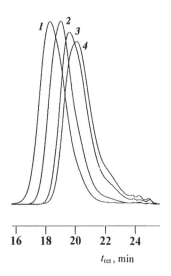

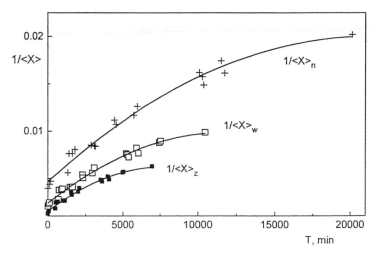

Figure 5. Time development of SEC records of the copolymer of phenylacetylene and p-nitrophenylacetylene degrading in THF at room temperature; curve − degradation time: **1** − 38; **2** − 1 620; **3** − 10 100; **4** − 20 130 min.

Figure 6. Kinetic dependences according to equations (3) to (5) for degradation of the copolymer of phenylacetylene and p-nitrophenylacetylene in THF at room temperature. T is degradation time t for $\langle X \rangle_n$, 1/3 of integral from equation (4) for $\langle X \rangle_w$, and 1/2 of integral from equation (5) for $\langle X \rangle_z$.

Poly(iodophenylacetylene)s (PIPA). Divers degradation behavior is observed for three positional isomers (2-, 3- and 4-) of PIPA prepared with WOCl$_4$-based catalyst (high-trans polymers). Oxidative degradation of 4-PIPA/W has been studied recently and found to suit the laws of polymer random degradation *(41)*. The determined value of the rate constant $\nu = 2.6 \cdot 10^{-6}$ min^{-1} is close to that found for PPA/W. Degradation behavior of the other two isomers is seen from Figure 7. The ortho isomer (2-PIPA/W) is virtually stable in aerated THF solution while 3-PIPA/W is unstable. The latter degrades obeying the laws of the random degradation without an increased formation of low-MW fragments. The value of the rate constant $\nu = 1.3 \cdot 10^{-6}$ min^{-1} determined for 3-PIPA/W is half of that found for 4-PIPA. This difference, together with stability of 2-PIPA/W clearly point to an importance of the ring-substituent position for a degradability of substituted poly(phenylacetylene)s.

It is notable that also many other poly(phenylacetylene)s with bulky substituents (e.g., -Si(CH$_3$)$_3$) in ortho position are quite stable in air *(40, 43)*. These polymers stand out by a high extent of π-conjugation, which is ascribed to the increased stiffness of their main chains due to the effect of bulky ortho substituents. They are supposed to render the plane of the phenyl rings perpendicular to the main chain axis making the main chain more planar and better conjugated. So it is also in the case of 2-PIPA/W that shows highest extent of π-conjugation of all PIPAs *(43)*.

Poly[4-(triisopropylsilylethynyl)phenylacetylene] (PTSEPA). SEC records of the high-trans low-cis polymer prepared with WOCl$_4$-based catalyst (PTSEPA/W) measured at various degradation times are shown in Figure 8. Unlike the case of PPA/W, the degradation of PTSEPA/W is accompanied by a formation of small but easily detectable amounts of oligomeric fragments, mainly of the monomeric type. It means that this degradation is not of the purely random type. Corresponding kinetic plots are shown in Figure 9. They exhibit initial nonlinearity (lasting five days) followed by a linear course on a long time interval. This behavior can be ascribed to a presence of weak links at which the degradation takes place preferably *(28)*. Values of ν determined from linear parts of kinetic dependences are as follows: $\nu_n = 2.1 \cdot 10^{-7}$ min^{-1}, $\nu_w = 1.25 \cdot 10^{-7}$ min^{-1} and $\nu_z = 1.1 \cdot 10^{-7}$ min^{-1}. The degradation of high-cis PTSEPA/Rh sample is also accompanied by a formation of low-MW byproducts. However, the kinetic plots for this sample are linear and provide values of the rate constant: $\nu_n = 3.6 \cdot 10^{-6}$ min^{-1}, $\nu_w = 2.0 \cdot 10^{-6}$ min^{-1} and $\nu_z = 1.7 \cdot 10^{-6}$ min^{-1} that are about one order of magnitude higher compared to those for PTSEPA/W. This suggests that cis units

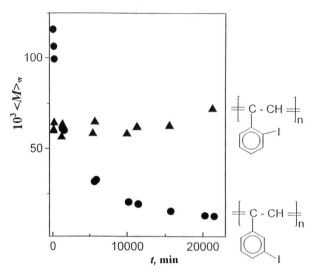

Figure 7. Time dependences of $\langle M \rangle_w$ measured for solutions of 2-PIPA/W (a) and 3-PIPA/W (b) in THF exposed to air at room temperature (both samples prepared with $WOCl_4/Ph_4Sn$ catalyst in benzene/dioxane, 1:2).

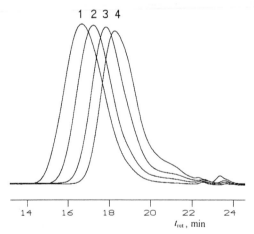

Figure 8. Time development of SEC records of PTSEPA/W sample (prepared with $WOCl_4/Ph_4Sn$ catalyst in benzene/dioxane, 1:2) degrading in THF at room temperature; curve − degradation time: **1** − 37; **2** − 3 020; **3** − 7 530; **4** − 12 930 min.

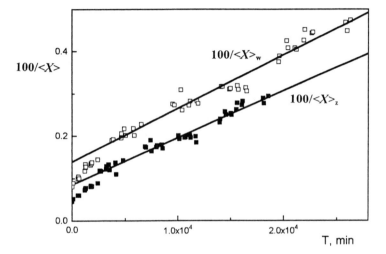

Figure 9. Kinetic dependences according to equation (4) and (5) for the degradation of PTSEPA/W sample in THF at room temperature. T is 1/3 of integral from equation (4) for $\langle X \rangle_2$, and 1/2 of integral from equation (5) for $\langle X \rangle_z$.

present in the PTSEPA/W can be the weak points that are responsible for initial faster degradation of this polymer.

Mechanistic Considerations. Characteristic features of the oxidative degradation of poly(phenylacetylene)s can be summarized as follows.

- Degradability of isomeric poly(phenylacetylene)s is strongly influenced by the ring-substituent position increasing from ortho to para isomer.
- Poly(phenylacetylene)s that undergo oxidative degradation degrade without any induction period.
- The trans-rich poly(phenylacetylene)s virtually degrade in accordance with laws of the random polymer degradation.
- Degradation of cis-rich poly(phenylacetylene)s can be described as the random cleavage of macromolecules into fragments that consecutively undergo a partial chain-end degradation (elimination of cyclotrimers and monomeric type species).

These observations suggest that the oxidative degradation of substituted polyacetylenes is a two step process including (Figure 10):

1. the random-type oxidative cleavage of macromolecules to two fragments with energetically excited newly formed ends; and
2. consecutive relaxation of excited fragment ends taking place either in a nonreactive and/or reactive way.

The absence of induction period of oxidative cleavage is explained by a presence of free spins (delocalized unpaired electrons) on macromolecules of substituted polyacetylenes that has been clearly demonstrated by numerous ESR measurements, see, e.g., Refs *(24, 26, 46 - 48)*. As there is no spin prohibition of the reaction between the triplet oxygen and free spins, the autoxidation of substituted polyacetylenes can easily start without any induction period. Random distribution of free spins along main chains explains randomness of the process ad a).

Resistance of ortho-substituted poly(phenylacetylene)s to oxidative degradation can be tentatively explained by crowding of main chains by ortho substituents. In addition, the concentration and energy of free spins should be lowered in these polymers due to higher extent of π-conjugation of their stiff macromolecules.

The nonreactive relaxation of primary fragment ends can proceed as the excitation energy transfer along the conjugated main chain and, finally, to solvent molecules (Figure 10, process 2a). Good conjugation of main chains typical of the trans-rich polymers should facilitate this type of relaxation. Both vibrational and electronic excitation of fragment ends can be expected. However, a chemiluminescence was not detected during the degradation of poly(phenylacetylene)s. Most probably, this is owing to high efficiency of the non-radiative transfer of excitation energy in their macromolecules because also the photoluminescence of these polymers is virtually negligible.

Partial reactive relaxation of ends of cis-rich macromolecular fragments, i.e., the elimination of chain-end units or oligomeric fragments like cyclotrimers (Figure 10, process 2b) is in agreement with lower extent of π-conjugation in cis-rich macromolecules. Actually, the cis units act as weak π-conjugation defects such that the energy delocalization in conjugated main chains is easier in trans as compared to cis polyacetylenes *(24, 49, 50)*. Lowered extent of π-conjugation should lead to a decrease in the non-reactive relaxation and increase in the probability of chemical transformation of an

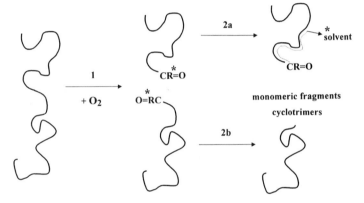

Figure 10. Scheme of suggested two-step general mechanism of the oxidative degradation of polyacetylenes: **1** Random-type oxidative cleavage of a macromolecule to two fragments with energetically excited newly formed chain ends; **2a** nonreactive and **2b** reactive relaxation of excited fragment ends.

excited fragment end. In addition, the cis configuration of main chains should favor a formation of cyclotrimers.

Influence of Polymer Degradation on Results of SEC Analysis

The degradation rate constant v appearing in equations 3 to 5 is, actually, the first-order rate constant of a main chain bond decay in the process of random polymer degradation *(33 -36)*. It is thus obvious that from a value of v the half-time of a degrading macromolecule with DP equal to X, $\tau(X)$, can be easily calculated: $\tau(X) = \ln2/v(X-1)$. Taking the value of v $= 2 \cdot 10^{-6}$ min^{-1} found for PPA/W in stabilized THF *(16)* and assuming an autoxidatively degrading macromolecule of $X = 5\,000$, then the macromolecule half-time $\tau(X)$ is about 70 minutes only. Within that time just one half of original macromolecules remain intact the other being disrupted to fragments.

The calculated time interval is well comparable with the time needed for executing a polymer SEC analysis. Therefore, the following question has arisen: does the autoxidative degradation take place inside SEC columns during the analysis of an oxygen-sensitive polymer performed with an eluent that is not free from air?

Experimental Evidence for the Polymer Degradation During SEC Analysis.
The preceding question has recently been answered. Using three types of experimental approaches and PPA/W as the model polymer *(16)* has proved the degradation of oxygen-sensitive polymer inside SEC column:

a) Comparison of results of SEC analyses performed in THF saturated with argon and air, respectively. In the first experiment, a sample of PPA/W was, under the argon atmosphere, dissolved in THF saturated with argon. This THF was also used as eluent in the subsequent SEC analysis. The following values were obtained (MW relative to PS standards): $<M>_w = 475\,000$, $I_n = <M>_w/<M>_n = 1.62$. In the second experiment, the same PPA/W sample was dissolved in aerated THF and the SEC analysis was performed in aerated THF with the following result: $<M>_w = 405\,000$, $I_n = 1.70$. The drop in MW values that corresponds to the value of degradation rate constant $v = 2.2 \cdot 10^{-6}$ min^{-1} is in a good agreement with the value stated above for degradation in stabilized THF (the degradation time t consists of 20 min of PPA dissolving and ca 16 min of the SEC peak retention time).

b) Comparison of results of polymer SEC analyses performed at different eluent flow rates. In these experiments, another PPA/W sample has been analyzed at two different flow rates by using SEC-LALLS device and non-stabilized THF. The following values were obtained: $<M>_w = 850\,000$ at a flow rate of 2.5 cm^3 min^{-1}, (overall degradation time $t = 45$ min) and $<M>_w = 520\,000$ at a flow rate of 0.5 cm^3 min^{-1}, ($t = 100$ min). From these data, $v = 2.7 \cdot 10^{-6}$ min^{-1} was obtained, which is in a good agreement with the value mentioned above for PPA in non stabilized THF.

c) Comparison of results of SEC analyses performed with and without interrupting the eluent flow for a certain time. In these experiments, polystyrene standards (PL Bristol, GB), MW 2 050 000 and 18 100, were added into analyzed solutions as internal calibration standards (see Figure 11). Independence of the PS standards retention times on the flow interruption was examined in blank experiments. Then a sample of PPA/W ($<M>_w = 305\,000$) was dissolved in the THF solution of PS standards and resulting mixture was analyzed without the eluent flow interruption (curve *1*) to obtain the reference record. In the "key"

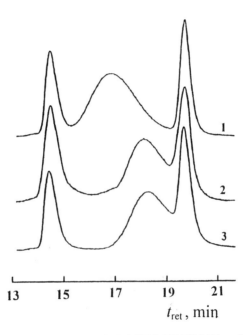

13 15 17 19 21

t_{ret}, min

Figure 11. SEC records of mixtures of PPA/W (MW 305 000) and two PS standards (MW 2 050 000 and 18 100): *1* fresh solution, common analysis; *2* fresh solution, the eluent flow interrupted 2.5 min after injection and the analysis finished after next 16 h; *3* solution aged in air for 16.5 h, common analysis. Reproduced from Collection of Czechoslovak Chemical Communication 61 (1996), J. Vohlídal et al., page 124, ref. *(16*, Copyright 1996, with kind permission of the Institute of Organic Chemistry and Biochemistry, Academy of Sciences of the Czech Republic, 166 10 Prague 6, Czech Republic.

experiment, a fresh solution of PPA and PS standards was injected to SEC system and the eluent flow was stopped after 2.5 min of the SEC analysis onset. The analysis was accomplished 16 h later (curve *2*). Finally, the stock solution of PPA and PS standards that was allowed in contact with air during the main experiment, was injected and analyzed without an interruption to acquire information enabling a comparison of extents of PPA degradation inside and outside SEC column (curve *3*).

Results of the performed experiments clearly show that: *(i)* PPA degrades inside the SEC columns during the analysis; and *(ii)* the rates of PPA degradation in open air and inside an SEC column are practically equal.

Model of SEC Analysis Accompanied by the Polymer Degradation in Columns.

Once the degradation of degradable polymer inside SEC columns had been demonstrated, two principal questions arose: *(i)* to what extent the polymer characteristics obtained from such measurements are distorted and how this distortion depends on the method of SEC trace evaluation; and *(ii)* whether the measured (and therefore distorted) values of DP averages $<X>_i^t$ (i = n, w, z), when treated according to equations 3 to 5, can provide correct values of the degradation rate constant v and original DP averages of non-degraded polymer. Answers to both these questions have recently been found as a result of the theoretical investigation and mathematical modeling of the process of SEC analysis accompanied by the random degradation of the analyzed polymer *(51)*.

Distortion of Measured DP Averages. The overall process of SEC analysis of polydisperse degradable polymer has been modeled by the set of continuity equations involving size-exclusion based partitioning, eddy diffusion (axial dispersion) and random degradation of the macromolecules passing the column. Numerical solution to the set of continuity equations has provided: *(i)* time dependences of the axial concentration profiles of X-mers visualizing a course of the separation process (see Fig. 12); *(ii)* simulated SEC records; and *(iii)* DP distribution of every SEC slice from which polydispersity of every polymer fraction could be calculated (see Fig. 13).

Table I. Comparison of $<X>_n$ and $<X>_w$ values obtained by various evaluation methods from SEC records simulated for various degradation rate constants v

	v, min^{-1}	calibration curve	LS detector	exact approach	equations 3 and 4
$<X>_n$	0	4 978	5 086	4 997	5 000
	$2.4 \cdot 10^{-6}$	4 800	4 756	4 548	4 568
	$2.4 \cdot 10^{-5}$	3 643	3 000	2 540	2 543
$<X>_w$	0	7 508	7 491	7 491	7 500
	$2.4 \cdot 10^{-6}$	7 335	7 050	7 050	7 052
	$2.4 \cdot 10^{-5}$	5 646	4 541	4 541	4 501

SOURCE: Adapted from ref. *(51)*, Copyright 1997, with kind permission of Elsevier Science - NL, Sara Burgerhartstraat 25, 1055 KV Amsterdam, The Nederlands.

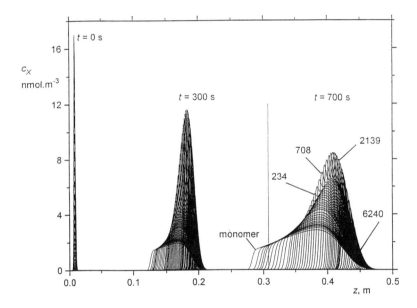

Figure 12. Time evolution of axial concentration profiles of selected X-mers at indicated times of SEC analysis simulated with the value of $\nu = 2.4 \cdot 10^{-5}$ min^{-1}. Reproduced from Journal of Chromatography A786 (1997), Z. Kabátek et al., page 214, ref. *(51)*, Copyright 1997, with kind permission of Elsevier Science – NL, Sara Burgerhartstraat 25, 1055 KV Amsterdam, The Nederlands.

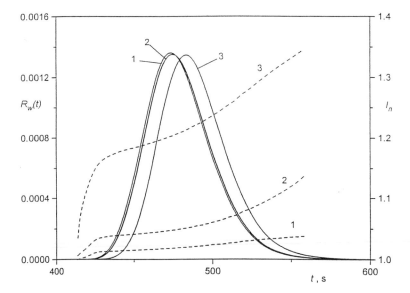

Figure 13. Simulated SEC records of a weight-concentration detector $R_2(t)$ (solid lines, left axis scale) and the polydispersity indices, $I_n = \langle X \rangle_{2,i}/\langle X \rangle_{n,i}$, of SEC slices (dashed lines, right axis scale). Curve − value of ν in min^{-1}: **1** − 0; **2** − $2.4 \cdot 10^{-6}$; **3** − $2.4 \cdot 10^{-5}$. Reproduced from Journal of Chromatography A786 (1997), Z. Kabátek et al., page 215, ref. *(51)* Copyright 1997, with kind permission of Elsevier Science − NL, Sara Burgerhartstraat 25, 1055 KV Amsterdam, The Nederlands.

Simulated SEC records were evaluated by three methods: (i) calibration curve method; (ii) light scattering (LS) detector method, that is, by using simulated data of an LS detector; and (iii) exact approach method, that is, by using knowledge of DP distributions of all SEC slices (that is inaccessible experimentally). The last method has provided the reference values of DP averages and, in addition, allowed to compare the obtained results with values calculated from equations 3 to 5 to verify reliability of simulated data.

The main results concerning the extent of the obtained DP average values' distortion are summarized in Table I. The DP averages obtained by the exact approach evaluation of simulated data agree satisfactorily with those calculated from equations 3 and 4, which proves a reliability of the used simulation methods. The LS-detector evaluation method yields correct values of $<X>_w^t$ corresponding to a given degradation time but overestimated values of $<X>_n^t$. This overestimation is expected because, when applying this method, we use values of weight average DP of every i-th SEC slice, $<X>_{w,i}$, for the calculation of $<X>_n$ value of given polymer. Perhaps the most worth noting fact is that the values of both $<X>_n$ and $<X>_w$ obtained by the calibration curve method are considerably overestimated, the relative increase being more pronounced for higher values of v. The axial concentration profiles shown in Figure 12 provide the explanation for this finding. It is evident that various low-DP fragments significantly contribute to the overall response of concentration detector belonging to the i-th SEC slice. However, the X_i value ascribed to the i-th SEC slice from the calibration curve belongs to the largest (that is yet non-degraded, original) X-mer molecules contained in the slice regardless the presence of smaller fragments. Therefrom it is obvious that both $<X>_n$ and $<X>_w$ values obtained by the calibration curve method must be overestimated.

Plausibility of Values of v, $<X>_n^0$ and $<X>_w^0$ estimated by using systematically distorted $<X>_n^t$ and $<X>_w^t$ values and equations 3 and 4 have been checked up by the computer experiment simulating a series of subsequent SEC analyses of a degrading polymer stock solution. The simulated SEC records were evaluated by all three methods like in the preceding case and the obtained simulated $<X>_n^t$ and $<X>_w^t$ values were treated according to equations 3 and 4. Resulting values of degradation rate constants v_n and v_w and initial DP averages $<X>_n^0$ and $<X>_w^0$ are summarized in Table II, together with the values

Table II. Values of DP averages of nondegraded polymer and rate constant v as obtained by treating simulated data according to Equations (3) and (4)

	calibration curve	LS detector	exact approach	input values
$<X>_n^0$	12280	6560	5000	5000
$10^5 \cdot v_n$, min^{-1}	2.46	2.25	2.41	2.40
$<X>_w^0$	10250	7330	7540	7500
$10^5 \cdot v_w$, min^{-1}	2.37	2.53	2.37	2.40

SOURCE: Adapted from ref. (51), Copyright 1997, with kind permission of Elsevier Science - NL, Sara Burgerhartstraat 25, 1055 KV Amsterdam, The Nederlands.

used as inputs for calculations (the correct, referent values). It should be emphasized here that presented results were obtained for an extremely low-stable model polymer the degradation rate constant of which is ca ten times higher than that of PPA/W. The discrepancy between the input and obtained values is, of course, smaller for more stable polymers.

Plausibility of measured v values. As it is seen from Table II, both $<X>_n^t$ and $<X>_w^t$ values, despite their systematic distortion, yield plausible value of the degradation rate constant v. The obtained particular v_n and v_w values are randomly scattered around the correct v value within a range that does not exceed usual experimental error. Surprisingly, the $<X>_n^t$ and $<X>_w^t$ values obtained by the calibration curve methods provide even better values of v as compared to those obtained by LS detector method. It can be shown by means of comparative quantitative analysis of values obtained by LS detector and exact approach methods that this is a consequence of the use of $<X>_{w,i}$ averages in the SEC records evaluation by LS-detector method. Nevertheless, the average of v_n and v_w values obtained from LS-detector data is very close to the correct value of v.

Accessibility of $<X>_n^0$ and $<X>_w^0$ values. The data summarized in Table II clearly show that only $<X>_w^0$ value can be estimated reasonably (with an acceptable error) provided the $<X>_w^t$ values obtained by the LS detector method are available. Found deviation of ca 2 % falls within the common experimental error of SEC method. On the contrary, the value of $<X>_n^0$ obtained by using the LS detector data is overestimated about 30 %, which significantly exceeds usual experimental error. However, in the case of more stable polymers, including PPA/W with ten times lower constant v, a determined value of $<X>_n^0$ might be acceptable. On the other hand, the $<X>_n^t$ and $<X>_w^t$ values obtained by the calibration curve method provide inapplicable values of $<X>_n^0$ and $<X>_w^0$. This is perhaps best demonstrated by the finding that so estimated $<X>_n^0$ value can be even higher than that of $<X>_w^0$ (Table II, first column), which is impossible event from the mathematics point of view. There is likely no reasonable way of eliminating this systematic error here. Therefore, a usage of the calibration curves method in determining the values of $<X>_n^0$ and $<X>_w^0$ should be avoided.

Conclusions

PPA and its meta and para ring-substituted derivatives degrade without any induction period when exposed to air at room temperature, whereas PPAs carrying bulky substituents in the ortho position are stable in air. This oxidative degradation can be regarded as a two-step process consisting of: (i) the random cleavage of main chains to primary fragments initiated by a spin-allowed reaction of triplet oxygen with delocalized free spins, followed with (ii) a relaxation of energetically excited fragment ends taking place in the non-reactive and/or reactive way. The former consists in the excitation energy transfer along main chains and, finally, its dissipation to surrounding molecules. The latter proceeds as the elimination of low-MW species from excited fragment ends.

Oxidative degradation of PPAs is slow in the solid state but fast in solutions. The degradation in solution is often so fast that it significantly affects the results of SEC analysis of an investigated polymer. This new phenomenon concerning the practice of SEC was experimentally demonstrated on high-trans PPA samples and its theoretical treatment has provided a deep insight into distortion of an SEC analysis results owing to the polymer degradation in columns. Computer experiments have shown that experimentally obtained, systematically biased values of $<M>_n^t$ and $<M>_w^t$ usually provide incorrect values of MW averages of non-degraded polymer but plausible values of the degradation rate constant v. This finding is of essential importance because, otherwise, values of the degradation rate constant could hardly be available for this type of polymers.

Despite the above complications, the SEC method is one of the most powerful tools for investigating the autoxidative degradation of substituted polyacetylenes because: (i) it is perhaps the most sensitive, ordinarily available method for detecting a polymer degradation even in its early stage; (ii) it provides information important for revealing the cleavage mode of macromolecules; and (iii) it makes easy accessible the time dependence of MW averages needed for the kinetic analysis of the degradation.

Acknowledgments. Authors are indebted to all collaborators who have contributed to their research in this field as co-authors of particular published and prepared articles (see References). Financial supports of the *Commission of EC* (PECO program, supplementary contract ERBCIPDCT940617) and the *Grant Agency of the Charles University* (189/97/B-CH) are gratefully acknowledged.

Literature Cited

1. Rooney J. G.; Ver Strate, G. In *Liquid Chromatography of Polymers and Related Materials III;* Cazes, J., Ed.; Marcel Dekker: New York, 1981; pp 207-235.
2. Hunt, B. J.; Holding, S. R. *Size Exclusion Chromatography;* Blackie: Glasgow, 1989.
3. Giddings, J. C. *Unified Separation Science;* Wiley: New York, 1991.
4. Slagowski, M. L.; Fetters, L. J.; McIntyre, D. *Macromolecules* **1974**, *7*, 394.
5. Kirkland, J. J. *J. Chromatogr.* **1976**, *125*, 231.
6. Huber, C.; Lederer K. H. *J. Polym. Sci.: Polym. Lett. Ed.* **1980**, *18*, 535.
7. Mei-Ling, Y.; Liang-He, S. *J. Liq. Chromatogr.* **1982**, *5*, 1259.
8. Giddings, J. C. *Adv. Chromatogr.* **1982**, *20*, 217.
9. Barth, H. G.; Carlin, F. J., Jr. *J. Liq. Chromatogr.* **1984**, *7*, 1717.
10. Grinshpun, V.; O'Driscoll, K. F.; Rudin, A. *J. Appl. Polym. Sci.* **1984**, *29*, 1071.
11. Halbwachs, A.; Grubišic-Gallot, Z. *Makromol. Chem. Rapid Commun.* **1986**, *7*, 709.
12. Barth, H. G. *ACS Symp. Ser.* **1987**, *352*, 29.
13. Lesec, J.; Lecacheux, D.; Marot, G. *J. Liq. Chromatogr.* **1988**, *11*, 2571.
14. deGroot, A. W.; Hamre, W. J. *J. Chromatogr. A* **1993**, *648*, 33.
15. Žigon, M.; The, N. K.; Shuyao, C.; Grubišic-Gallot, Z. *J. Liq. Chromatogr.* **1997**, *20*, 2155.
16. Vohlídal, J.; Kabátek, Z.; Pacovská, M.; Sedláček, J.; Grubišic-Gallot, Z. *Collect. Czech. Chem. Commun.* **1996**, *61*, 120.
17. *Nonlinear Optical and Electroactive Polymers;* Prasad, P. N.; Ulrich, D. R., Eds.; Plenum Press: New York, 1987.

286

18. *Conjugated Polymers;* Brédas, J. L.; Silbey, R., Eds.; Kluwer Academic Publishers: Dordrecht, 1991.
19. Long, N. J. *Angew. Chem. Int. Ed. Engl.* **1995,** *34,* 21.
20. *Photoactive Organic Materials;* Kajzar, F., Ed.; NATO ASI Series; Kluwer Academic Publishers: Dordrecht, 1996.
21. *Handbook of Organic Conductive Molecules and Polymers;* Nalwa, H. S., Ed.; Wiley: New York, 1996.
22. Manners, L. I. *Angew. Chem. Int. Ed. Engl.* **1996,** *35,* 1602.
23. Nešpůrek, S. In *Photoactive Organic Materials;* Kajzar, F., Ed.; NATO ASI Series; Kluwer Academic Publishers: Dordrecht, 1996; pp 411-430.
24. Chien, J. C. W. *Polyacetylene - Chemistry, Physics and Material Science;* Academic Press: New York, 1984; Chapter 7.5, pp 288.
25. Neoh, K. G.; Kang, E.; Tan, K. L. *Polym. Degrad. Stab.* **1989,** *26,* 21.
26. Vohlídal, J.; Rédrová D.; Pacovská, M.; Sedláček, J. *Collect. Czech. Chem. Commun.* **1993,** *58,* 2651.
27. Tabata, M.; Tanaka, Y.; Sadahiro, Y.; Sone, T.; Yokota, K.; Miura, I. *Macromolecules* **1997,** *30,* 5200.
28. Jellinek, H. H. G. In *Aspects of Degradation and Stabilization of Polymers;* Jellinek, H. H. G., Ed.; Elsevier: Amsterdam, 1978; Chapter 1, pp 1 - 38.
29. Abbas, K. B. In *Liquid Chromatography of Polymers and Related Materials II;* Cazes, J.; Delamar, X., Eds.; Chromatographic Science Series Vol. 13, Marcel Dekker: New York, 1979; pp 123-142.
30. Basedow, M.; Ebert, K. H.; Ederer, H. E. *Macromolecules* **1978,** *11,* 774.
31. Ballauf, M.; Wolf, B. A. *Macromolecules* **1981,** *14,* 654.
32. Netopilík, M.; Kubín, M.; Schultz, G.; Vohlídal, J.; Kossler, I.; Kratochvíl, P. *J. Appl. Polym. Sci.* **1990,** *40,* 1115.
33. Montroll, E.W.; Simha, R. *J. Chem. Phys.* **1940,** *8,* 721.
34. Simha, R. *J. Appl. Phys.* **1941,** *12,* 569.
35. Guilett, J. *Polymer Photophysics and Photochemistry;* Cambridge University Press: Cambridge, 1985.
36. Vohlídal, J. *Macromol. Rapid Commun.* **1994,** *15,* 765.
37. Sedláček, J.; Vohlídal, J.; Grubišic-Gallot, Z. *Makromol. Chem. Rapid Commun.* **1993,** *14,* 51.
38. Pfleger, J.; Nešpůrek, S.; Vohlídal, J. *Mol. Cryst. Liq. Cryst.* **1989,** *166,* 143.
39. Sedláček, J.; Pacovská, M.; Vohlídal, J.; Grubišic-Gallot, Z.; Žigon, M. *Macromol. Chem. Phys.* **1995,** *196,* 1705.
40. Shirakawa, H.; Masuda, T.; Takeda, K. In *The Chemistry of Tripple-Bonded Functional Groups;* Patai, S., Ed.; Supplement C2; Wiley: New York, **1994,** Chapter 17, pp. 945-1016.
41. Vohlídal, J.; Sedláček, J.; Pacovská, M.; Lavastre, O.; Dixneuf, P. H.; Balcar, H.; Pfleger, J. *Polymer* **1997,** *38,* 3359.
42. Sedláček, J.; Vohlídal, J.; Cabioch, S.; Lavastre, O.; Dixneuf, P.H.; Balcar, H.; Štícha, M.; Pfleger, J.; Blechta, V. *Macromol. Chem. Phys.* **1998,** *199,* 155.
43. Vohlídal, J.; Sedláček, J.; Patev N., Pacovská, M.; Cabioch S.; Lavastre, O.; Dixneuf, P. H.; Blechta, V.; Matějka P.; Balcar, H. *Collect. Czech. Chem. Commun.* **1998,** *63,* 1815.

44. Kishimoto, Y.; Eckerle, P.; Miyatake, T.; Ikariya, T.; Noyori, R. *J. Am. Chem. Soc.* **1994,** *116,* 12131.

45. Furlani, A.; Napoletano, C.; Russo, M.V.; Feast, J.W. *Polym. Bull.* **1986,** *16,* 311.

46. Tabata, M.; Sone, T.; Sadahiro, Y.; Yokota, K.; Nozaki, Y. *J. Polym. Sci. A, Polym. Chem.* **1998,** *36,* 217.

47. Tabata, M.; Sone, T.; Sadahiro, Y.; Yokota, K. *Macromol. Chem. Phys.* **1998,** *199,* 1161.

48. Tabata, M.; Kobayashi, S.; Sadahiro, Y.; Nozaki, Y.; Yokota, K., Yang, W. *J. Macromol. Sci., Pure. Appl. Chem.* **1997,** *A34,* 641.

49. Gussoni M., Castiglioni C., Zebri G.: In *Spectroscopy of Advanced Materials;* Clark R. J. H., Hester R.E., Eds.: Wiley: New York, **1991**; Chapter 5, pp. 251-353.

50. Wada, T.; Wang, L.M.; Okawa, H.; Masuda, T.; Tabata, M.; Wan, M.X.; Kakimoto, M.A.; Imai, Y.; Sasabe, H. *Mol. Cryst. Liq. Cryst.* **1998,** *294,* 245.

51. Kabátek, Z.; Gaš, B.; Vohlídal, J. *J. Chromatogr. A* **1997,** *786,* 209.

Chapter 20

Analysis of Polysaccharides by SEC[3]

D. T. Gillespie[1] and H. K. Hammons[2]

[1]Chevron Chemical, LLC, 1862 Kingwood Drive, Kingwood, TX 77339
[2]Viscotek Corporation, 15600 West Hardy, Houston, TX 77060

Polysaccharides (PSC) are ideally suited for SEC[3] analysis, which consists of three detectors (Refractometer, Viscometer, Light Scattering) coupled to a standard SEC/GPC system[1]. The overall approach and advantages of SEC[3] for molecular weight and structural determination have been described elsewhere[2].

The RI detector is necessary because PSC polymers do not have useful absorption in the UV.

The Viscometer is useful because many of these polymers are branched and the degree of branching can be measured through the intrinsic viscosity.

Light Scattering permits the direct measurement of molecular weight without the use of calibrants.

In this work, we demonstrate the application of SEC[3] to several representative PSC polymers, including dextran, maltodextrin, starch, carrageenan, hyaluronic acid, and chitosan. Most PSC's can be successfully analyzed in simple aqueous mobile phase solvents. Higher molecular weight starch seems to dissolve better in DMSO.[3] The extent and path of degradation in various conditions is of particular importance for PSC's used in food processing and that can be done quite accurately using SEC[3] to measure the distribution of both molecular weight and branching.

The micro-structural properties of PSC systems can be examined by determining the bulk relationships between molecular weight and molecular density within the macromolecules. Moreover, as a PSC grows with respect to repeat unit, its hydrodynamic volume (dL / molecule) will change depending upon the consistency of density across the 3-dimension growth pattern. Traditionally, intrinsic viscosity (dL / g) is used as a direct measure of hydrodynamic volume (HV). Since HV is the reciprocal of the molecular density of a solvated molecule, the measurement of intrinsic viscosity can be useful in monitoring growth mechanisms. In order to evaluate changing growth patterns as a function of chain length, Size Exclusion Chromatography, SEC, is used to provide a hydrodynamic size separation for the macromolecules.

Background

The Triple Detector System used in conjunction with size exclusion chromatography consisting of an on-line molecular weight, density, and concentration specific detector has been referred to as an SEC^3 system[4]. Typically, the molecular weight specific detector is a light scattering detector, although analogous substitutions can be made such as employing a viscometer with a Universal Calibration curve[5]. The density detector is typically a viscometer detector, and the concentration detector is typically a refractometer or UV detector. The measurement of polymer hydrodynamic volume is made by the coupling of the viscosity and molecular weight detectors and translated to radius of gyration through the following relationship[6]:

$$[\eta] = \Phi <r^2>^{3/2} / M \qquad (1)$$

Where Φ is the Flory viscosity constant, $<r^2>$ is the mean square end-to-end distance, $[\eta]$ is the intrinsic viscosity, and M is the molecular weight. The Flory constant may be modified by the Ptitsyn Eizner equation[7] to consider the overall molecular geometry as determined from the Mark-Houwink slope (a) by the following equation:

$$\Phi = 2.86 \times 10^{21}(1 - 2.63\varepsilon_j + 2.86\varepsilon_j^2) \qquad (2)$$

Where $\varepsilon_j = 1/3(2a_j - 1)$.

Polymer size, determined as the radius of gyration, Rg, may also be determined through multi-angle light scattering techniques[8]. However, this determination relies on the angular dis-symmetry function as related to the differences in relative light scattering intensities at multiple angles. For polysaccharides containing low molecular weight fractions, the differences between angles may become small compared to the light scattering detector noise at the lowest angle being measured[9]. In these cases, molecular weight may still be obtainable from the absolute light scattering intensity, even though the size measurement may not[10]. The SEC^3 technique relies on the absolute light scattering and viscometer intensities for its direct hydrodynamic size measurement and is constrained only by the full-scale signal at each chromatographic slice.

Universal Calibration is a well-proven means of molecular weight determination for systems obeying true size-exclusion mechanisms[11], and may be preferable in many circumstances that are not readily applicable to direct light scattering measurement such as determinations of samples with low molecular weights and low dn/dc values. However, non-size exclusion effects which can occur in aqueous systems[12] makes Universal Calibration less attractive than light scattering for the analysis of polysaccharides.

Experimental

The SEC[3] flow configuration for all of the experiments was a Waters HPLC dual piston pump, an Eppondorf column oven, which contained column sets that varied according to experimental conditions. The detection system was a RALLS light scattering detector followed by viscometer and laser RI in parallel configuration from Viscotek. The viscometer - laser RI combination along with the column oven was operated at 38°C. Sample introduction was done with a Micromeretics autosampler equipped with a fixed loop injector. The injection size typically was set at 50uL per 30cm column. The inter-detector volume was measured by injecting a narrow polyethylene glycol (PEG) standard of 50,000 molecular weight obtained from American Polymer Standards. This standard was also used to verify dn/dc, viscosity, and molecular weight calibration constants for each of the detectors.

For each set of polysaccharides, a dn/dc was calculated based on injected mass and the area of laser RI. It is important to note that to obtain proper dn/dc values for many polysaccharides, the samples were first dried overnight in a heated vacuum oven. Most of the polysaccharides were found to absorb 5-10% moisture. If this is not accounted for dn/dc calculations will be in error by an inverse amount. Since light scattering measurements are proportional to $(dn/dc)^2$, the errors introduced will be substantial in terms of absolute molecular weight. The dn/dc determined for the polysaccharides ranged from 0.142 to 0.150, depending upon the sample type.

The mobile phase buffer for all systems was $0.05M$ sodium nitrate, except for the chitosan work, which contained mixtures of sodium acetate and acetic acid depending upon the polarity of the mobile phase desired. Some pectin work was completed in pure water and at $0.01M$ sodium nitrate to examine molecular interaction with the mobile phase. Finally, work on some starches was in $1M$ sodium hydroxide and in DMSO with 0.03% lithium bromide. Flowrate was generally set at 0.8mL/min.

For Figures 1-3, 2 30-cm TSK PWxL columns were used in series. A third TSK PWxL column was added for the work done in Figure 4-9. Figures 10 and 11 were using just 1 column. The broad pullulan shown in Figure 3 was obtained by mixing 5 narrow pullulan standards together obtained from Polymer Laboratories. The dextran of nominal 500,00 MW was obtained from Pharmacia. Pectin samples were donated from the USDA, ARS, ERRC, as well as Aldrich Chemical Company. The batch triple detector experiment in Figure 12 was done with a 2.5mL non-porous glass bead column manufactured by Viscotek. The chitosan sample in Figures 12-14 was obtained from Sigma Chemical Company. One TSK PWxL column was used in Figures 13-17. The chitosan samples in Figures 15-17 were obtained from Ehwa Woman's University of Korea. Figures 18-20 used two TSK PWxL columns. The Carrageenan was obtained from Fluka, and the Hyaluronic Acid was obtained from Aldrich. The DMSO work depicted in Figures 21-23 used an American Polymer

Standard 15um mixed bed column. Flowrate was established at 0.3mL/min to minimize shear degradation of the high molecular weight portions. The detector temperatures were set at 60C and the column set was heated to 85C. The starch samples were obtained from the USDA, ARS, NRRC and the pullulan polysaccharides were obtained from Polymer Laboratories. The comparison in Figure 23 was made to 0.05M sodium nitrate and a single TSK PWxL aqueous system. The starch in Table III and Table IV was obtained from the USDA, ARS, NRRC. The mobile phase was 1M sodium hydroxide and the column set was a 60cm 15um column from Jordi Associates.

Examination of Detector Responses for Polysaccharides

A pullulan PSC mixture was injected on the SEC^3 system for the purpose of comparing the detector response factors. The mixture consists of a pullulan 400,000 molecular weight at 0.2 mg/mL and a pullulan 10,000 molecular weight at 0.4mg/mL injected concentration. The resulting chromatogram (Figure 1) illustrates the difference in signal intensities of the detectors. The refractometer is proportional to the concentration of the PSC, the viscometer is proportional to the product of intrinsic viscosity and concentration, and the light scattering detector is proportional to the product of molecular weight and concentration. A dextran standard (Figure 2) with a broad molecular weight polydispersity shows a noted detector skewness as the light scattering and viscometer detectors respond more strongly to the higher molecular weight and viscosity fractions. The dextran does not increase as rapidly in viscosity as do the previous pullulan standards. This phenomenon is caused by the higher density of the dextran molecules versus the linear pullulan, due to the addition of long chain branches. The addition of fairly regularly spaced shorter chain branches causes the overall intrinsic viscosity to be lower throughout the entire molecular weight distribution, while the addition of longer chain random branches causes the intrinsic viscosity to fall much faster across the high molecular weight distribution. This is directly observable in an overlay of a Mark-Houwink plot of a broad dextran and pullulan standard (Figure 3).

Comparison of Polysaccharide Structure

Maltodextrin (Figure 4), high-amylose starch (Figure 5), and corn syrup solids (Figure 6), are examples of PSC with very large polydispersities commonly depicted by the differential molecular weight distribution (Figure 7) directly obtainable from the SEC^3 method. The three samples (as well as the previous dextran) were run in 0.05M sodium nitrate. A structural overlay shown by the Mark-Houwink plot (Figure 8) ranks the amount of branching, at the same molecular weight, in these samples from lowest to highest in the following order:

> Dextran
> High Amylose Starch
> Maltodextrin
> Corn Syrup Solids

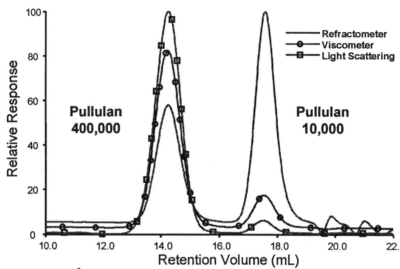

Figure 1. SEC^3 Chromatogram of Narrow Pullulan Mixture

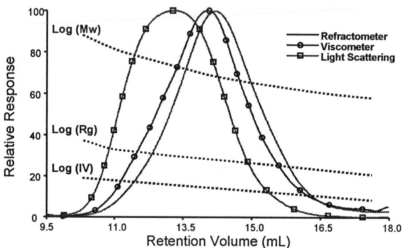

Figure 2. SEC^3 Chromatogram of Broad Dextran Mixture

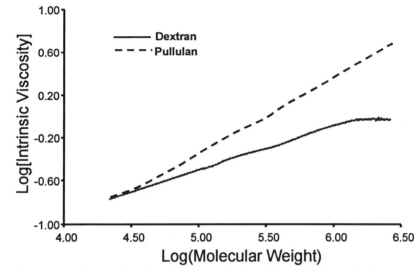

Figure 3. Mark-Houwink Overlay of Branched and Linear Polysaccharides

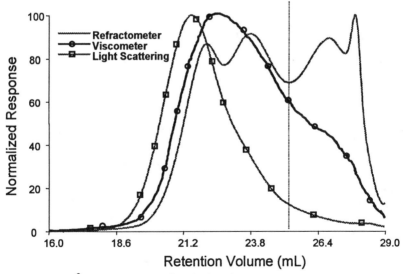

Figure 4. SEC³ Chromatogram of Maltodextrin

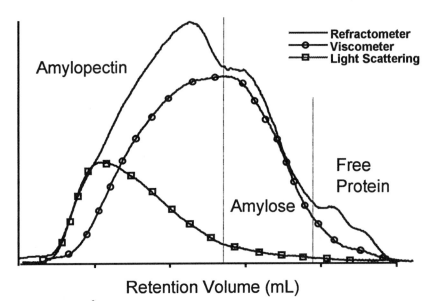

Figure 5. SEC³ High Amylose Starch

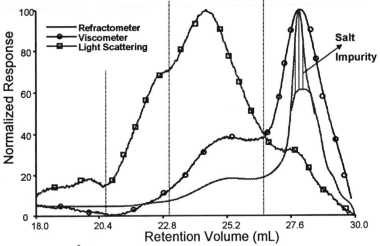

Figure 6. SEC³ Chromatogram of Corn Syrup Solids

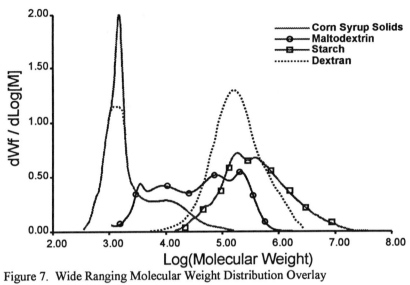

Figure 7. Wide Ranging Molecular Weight Distribution Overlay

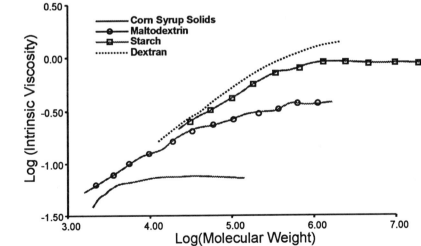

Figure 8. SEC³ Mark-Houwink Structural Comparison

It is evident that in the high amylose starch, there is a distinct break in the Mark-Houwink plot. Starch consists of two major components, linear amylose, and branched amylopectin.[13] The Mark-Houwink plot for the starch readily shows the difference between the amylose and the amylopectin fractions. There is incomplete chromatographic resolution between the species so the resultant plot is a curve (linear and branched material[14] co-eluting at the same hydrodynamic volume) rather than two distinct lines. However, it is clear that both the lower half and the upper half can be readily fit to a straight line for the estimation of the slope of each fraction (Mark-Houwink a value). Since both of these molecular fractions have similar characteristic repeat units[15], the Mark-Houwink intercept value (Log K) should be obtained by fitting the lowest molecular weight species only (amylose). This value can then be applied to calculate the branching within the higher molecular weight amylopectin fraction. It should also be noted that the amylopectin fraction has a slope that approaches zero. This strongly suggests that the volume of the molecule is expanding as a direct function of the mass which is present in a spherical model[16] (the amylopectin fraction is approaching a sphere).

It is important to note that the interpretation of triple detector results in general will not include why an overall shape change is occurring. Rather, the results must be combined with information about the general nature of the macromolecule. For example, three competing explanations for the behavior in the previous starch example are:

> The high molecular weight amylopectin is not fully solvated and the molecules are not swelling properly in this mobile phase.

> There is heavy branching within the amylopectin fraction.

> The amylopectin is aggregating (molecules are sticking together).

Some starch fractions did not solubilize fully in $0.05M$ sodium nitrate. Therefore, all three mechanisms are likely. In such a case, studies should be made to plot results as a function of concentration, molarity of the mobile phase, and type of mobile phase. For example, aggregation should be less of a problem at lower concentrations. Substituting the mobile phase with DMSO may solve the solvation problem. Changing the salt to $1M$ sodium hydroxide will also change the solubility (but may also yield limited sample life due to probable hydrolysis of the backbone and branches.)

Environmental Considerations in Polysaccharide Analysis

When PSC's are dissolved in aqueous buffers, several effects can occur including polyelectrolyte expansion[17], aggregation[18], and de-branching[19]. The first study involves the polyelectrolyte expansion of a solution of pectin in neutralized form.

The Mark-Houwink plot (Figure 9) examines the effect of three different buffer concentrations ranging from $0.05M$ sodium nitrate to pure water. The effect of reducing the salt concentration in the mobile phase greatly effects the expansion of the pectin molecules. This is evident by the dramatic slope change of the Mark-Houwink plot, indicating that the structure is elongating from a true random coil type molecule[20] to more of a rod-like structure due to lack of counter-ions in the mobile phase. Note that although the molecule becomes more rod-like, a true rod is not obtained, as there will always be some flexibility within the molecule. It should be noted that intrinsic viscosity must therefore be a function of the ionic strength of mobile phase buffer. However, it should be noted that the hydrodynamic volume is also a function of size as can be seen readily from the overlay of the RI chromatograms. Furthermore, this implies that the calculated radius of gyration, which is a function of the overall molecular volume, also varies as a function of the molarity of the mobile phase. Pectin in the protonated form (Figure 10) can also aggregate depending upon sample preparation. A Mark-Houwink plot overlaid with a linear reference and a molecular weight distribution plot (Figure 11) is a useful qualitative tool for monitoring the population of aggregated material within a given sample. Mark-Houwink plots can be overlaid and compared for overall aggregation content.

Chitosan solubility is a function of the pH of the mobile phase. A chitosan sample that was dissolved at a pH of 4.3 was tested for solubility across different pH ranges. As the pH was increased to 5.8, the chitosan precipitated. The intrinsic viscosity of chitosan was measured as a function of pH to examine the relationship between molecular folding and intrinsic viscosity. The chromatography columns were removed from the system and a 5cm column filled with non-porous, silanized, glass beads was used to retain the sample. The glass bead column was chosen for two major reasons, it provides good peak shape (normal distribution versus exponential separation on tubing alone) and it provides adequate sample dilution (necessary for dilution and numerical integration). Additionally, the batch holdup column is resistant to fast solvent changes and has a total volume of approximately 3-mL. The batch triple detector analysis (Figure 12) gives no separation, but yields weight-average molecular weight, molecular density, and molecular size results in under 5 minutes. The batch results (Table I) at several different pH conditions illustrate the contraction of the molecule prior to precipitation. At these measurements close to infinite dilution, the intrinsic viscosity (at any given molecular weight) readily indicates changes in molecular swelling.

Table I. Intrinsic Viscosity Relationship with pH of Chitosan

Mobile Phase pH	IV at $25^{\circ}C$
2.7	12.37
4.3	10.50
5.4	10.27

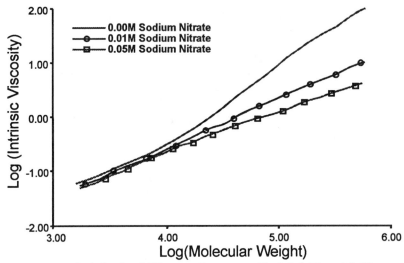

Figure 9. Pectin Solvation Differences Represented by Mark-Houwink Plot

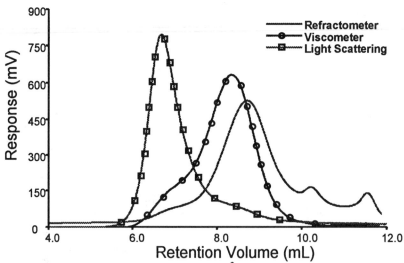

Figure 10. Pectin Aggregation as seen in SEC[3]

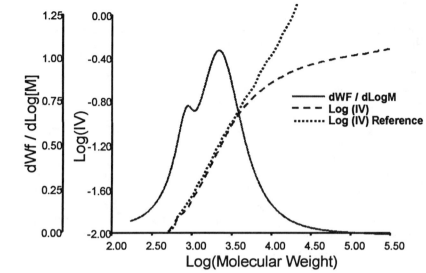

Figure 11. Mark-Houwink Plot of Pectin with Suspected Aggregates

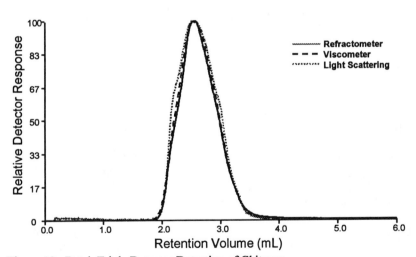

Figure 12. Batch Triple Detector Detection of Chitosan

Chilling of the chitosan also caused precipitation. An experiment was run at two equilibration temperatures. Chitosan was equilibrated for 24 hours at a sub-ambient temperature (5°C) and an elevated temperature (60°C). The chitosan was then injected directly on the chromatographic system (Figure 13). Surprisingly, the effects of molecular folding were still evident after several minutes when the sample eluted through the detector at ambient temperature. Some material was lost as shown by the decrease in refractive index area. However, the ratio of the light scattering area and viscometer area to the refractometer area yields the molecular weight and intrinsic viscosity, respectively. The decrease in both molecular size (retention volume) and intrinsic viscosity (IV/RI) is immediately evident in the sample pretreated at 5°C. It should be noted that the light scattering detector shows a signal consistent with a constant molecular weight (LS/RI has an equivalent area) for both the sample pretreated at 60°C and the sample pretreated at 5°C. The Conformation Plot (Figure 14) also shows the molecular contraction in terms of radius of gyration decrease at a constant molecular weight for the chitosan pretreatment at 5°C. The exact mechanics of this molecular contraction and expansion were not investigated. The batch triple detector experiment was then run at several 24 hour pretreatment temperatures to examine the temperature dependence of heat pretreatment of intrinsic viscosity (Table II). It is clear that the intrinsic viscosity shows an upward trend with temperature as the molecules may begin to unfold to a greater degree.

Table II.   Intrinsic Viscosity Relationship with Heat Pretreatment of Chitosan

Pretreatment °C	IV at 25°C
5	7.21
25	10.15
60	10.50
100	11.85

The aggregation phenomenon in chitosan is also a function of chitosan type and preparation[21]. The triple detector chromatogram easily distinguishes an aggregated chitosan (Figure 15) from a non-aggregated chitosan (Figure 16). It should be noted that since chitosan is a linear molecule[22], branching is ruled out as a possible structural change mechanism. However, in the extreme case of a large amount of aggregation, the Mark-Houwink plot (Figure 17) gives valuable insight into the mechanism itself. The Mark-Houwink plot of the aggregated chitosan actually produces a negative slope at the high end of molecular weight. This is generally not seen with most long-chain branching models which predicts that intrinsic viscosity will always increase as a function of molecular weight[23] (although the increase will approach zero at high branching frequencies and high molecular weights). Exceptions to this general rule are possible in dendritically branched and hyper-branched molecules. The negative slope that is found in the Chitosan plot indicates that the molecule is expanding faster in density than in radial growth. When molecules "stick" to each other, a spherical agglomeration encompassing a single hydrodynamic volume is created rather than each molecule forming its own

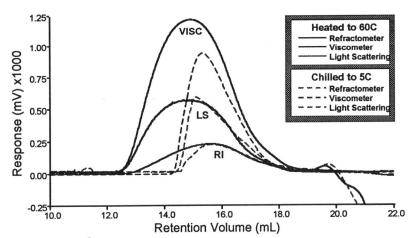

Figure 13. SEC³ Comparison of Pre-Treated Chitosan

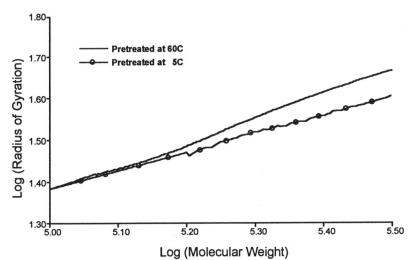

Figure 14. SEC³ Mark-Houwink Comparison of Pre-Treated Chitosan

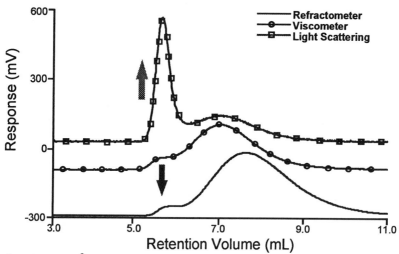

Figure 15. SEC^3 of Chitosan Containing a High MW Aggregate

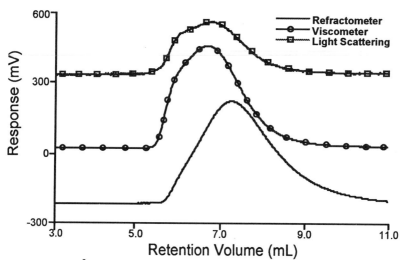

Figure 16. SEC^3 of Chitosan without a High MW Aggregate

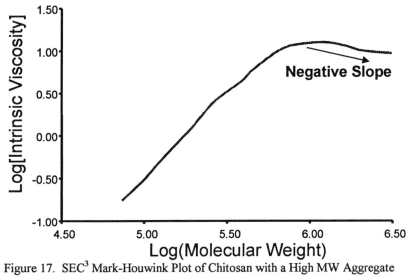

Figure 17. SEC^3 Mark-Houwink Plot of Chitosan with a High MW Aggregate

hydrodynamic volume. Thus, the overall density in solution is increased at approximately a proportional rate to the rate of aggregation.

Molecular Weight Determination at the Column Exclusion Limit

The definition of molecular density can be referred to as backbone molecular weight per repeat unit volume. When comparing random coil molecules, the density can be expressed as the relationship between backbone molecular weight per repeat unit length. Carrageenan (Figure 18) and hyaluronic acid (Figure 19) are linear molecules possessing very low molecular densities[24, 25] and both are shown to be random coil type molecules in aqueous buffers. This implies that the molecules will elute at very early retention volumes because they have a large hydrodynamic volume to molecular weight ratio. As the molecules approach the upper limit of the size exclusion column, separation no longer occurs. This is evident in the raw chromatogram of the carrageenan, which shows very low detector skewness from the light scattering and viscometer in reference to the refractometer. The differential radius of gyration distribution (Figure 20) has a very sharp rise in front, indicating that the column efficiency is inadequate for all of the molecular sizes contained in this molecule. This sharp rise is caused by a high population density, which is calculated as possessing the same radius (not being separated into a distribution). However, weight-average molecular weight values, weight-average intrinsic viscosity values, and weight-average radius of gyration values may be obtained across this excluded section of the chromatogram. This is analogous to the batch triple detector measurement discussed in the chitosan work.

Mobile Phase Considerations in the Preparation of Starches

Starch solubility is a problem because of the very high molecular weights of the molecules. Furthermore, the intense branching on starch samples with high amylopectin content also affects the rate of solvation of the molecules[26]. A concentration normalized comparison (Figure 21) of a dent starch and a pullulan PSC of 850,000 molecular weight (both prepared and run in DMSO) illustrates the signal differentiation of the detectors. The refractometer shows that the pullulan PSC has a very narrow hydrodynamic volume polydispersity while dent starch has a very broad elution profile. The viscometer demonstrates that the starch sample has an overall lower intrinsic viscosity than the pullulan (although its molecular weight is approximately 200 fold higher). The molecular weight difference is clearly detected by light scattering, which shows an extremely large peak height difference. It should be noted, however, that the light scattering detector has a sensitivity disadvantage at retention volumes past 6.5 mL. This disadvantage comes from the decreasing molecular weights past 6.5 mL coupled with the poor dn/dc of polysaccharides in DMSO (approximately 0.06). The viscometer signal (which is not proportional to dn/dc^2 and which retains stronger signal for linear macromolecular fractions) does not experience the sensitivity problem nearly as much. Although raising the concentration could increase the detector signals, this would cause solubility problems and separation problems on the high molecular

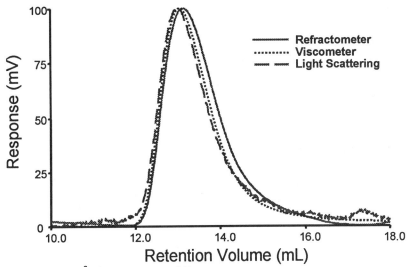

Figure 18. SEC³ Chromatogram of Carrageenan

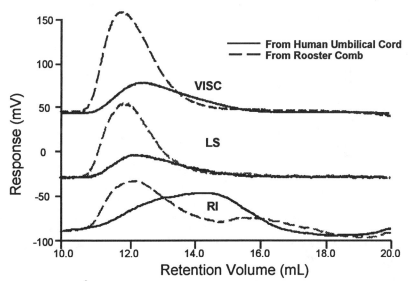

Figure 19. SEC³ Chromatogram of Hyaluronic Acid

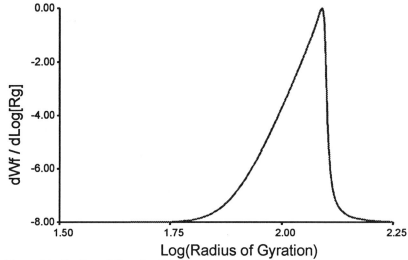

Figure 20. Radius of Gyration Distribution of Hyaluronic Acid

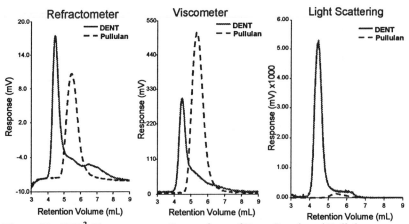

Figure 21. SEC^3 Chromatogram Comparison of Dent Starch and Pullulan

weight fractions. It is important to note this limitation of light scattering on ultra-broad molecular weight distributions.

The starch sample could be dissolved in DMSO and then run in an aqueous system (with a very large solvent impurity peak). This procedure can be shown to increase the upper end molecular weight distribution from a potato starch sample prepared in an aqueous buffer (Figure 22). However, the resultant downward trend of the high molecular weight fractions in the Mark-Houwink plot (Figure 23) uncovers an inherent solubility problem across the ultra-high molecular weight region as compared to similar work in DMSO. Starch could also be dissolved and run in sodium hydroxide. However, the branches and even the backbone may undergo hydrolysis. A Starch sample was prepared in $1M$ sodium hydroxide and injected onto the system. Subsequent injections show that the molecular weight deteriorates rapidly over time (Table 3). However, the intrinsic viscosity rises slightly over the same span. The conclusion is that the hydrolysis is preferentially affecting the branches of the molecule because the molecular weight and the molecular density are simultaneously being reduced (The molecule is less dense with the absence of the branches so the intrinsic viscosity can rise even though molecular weight has decreased.) The intrinsic viscosity is still very low considering the high molecular weights for the starch. This leads to the conclusion that not all of the branches have hydrolyzed, and, therefore, the hydrolysis occurs over a long time. After a longer period, both the molecular weight and the intrinsic viscosity begin dropping together (Table 4). This now can be explained by the molecular backbone also breaking as a function of the hydrolysis process. By 24 hours, a very low molecular weight, compared to the starting material, is observed by the triple detector.

Table III. Starch Degradation in NaOH – First 100 minutes

Minutes	20	60	80	100
Mw	218,000,000	183,000,000	160,000,000	140,000,000
IV	1.485	1.579	1.641	1.651

Table IV. Starch Degradation in NaOH – Over 24 Hours

Minutes	100	360	1440
Mw	160,000,000	40,000,000	200,000
IV	1.641	0.821	0.090

Conclusions

SEC^3 is a powerful technique to determine not only the molecular weight, but also the microstructure of polysaccharides. As the molecular weight of polysaccharides increases, the growth pattern may become linear, branched, or even aggregated. The addition of the viscometer detector allows these changes to be measured and

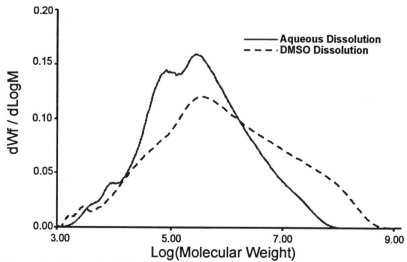

Figure 22. Molecular Weight Distribution Comparison of Potato Starch Dissolution Technique

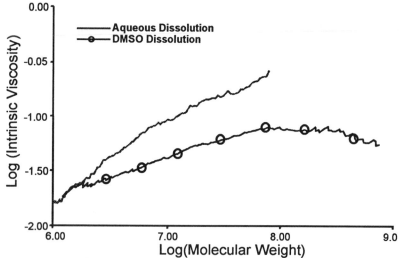

Figure 23. SEC³ Mark-Houwink Comparison of Potato Starch Dissolution Technique

compared between samples. Furthermore, measurements can be made showing differences as a function of sample preparation techniques and running conditions. The triple detector concept is useful in cases where there is little or no chromatographic resolution, as it can still yield weight-average molecular weight, weight-average intrinsic viscosity, and weight-average size information. With triple detector analysis operating in batch mode, running conditions can be varied quickly and sample throughput can be increased allowing timely studies of sample conditioning. Triple detector analysis is particularly useful for analyzing solubility problems and macromolecular reactions to environmental stresses. The usefulness of SEC^3 has distinct advantages even for systems using multi-angle light scattering detection and Universal calibration because it, in general, possesses additional structural sensitivity at low molecular weights and it can still provide structural information across regions of non-separation. Furthermore, when combined with multi-angle light scattering, it provides independent additional confirmation of structural changes occurring within polysaccharide samples. When combined with Universal Calibration techniques, it provides independent additional confirmation of molecular weight distribution within polysaccharide samples.

Literature Cited

1. M. A. Haney, C. Jackson, & W. W. Yau, "SEC-Viscometry-Right Angle Light Scattering (SEC-Visc_RALS)" *International GPC Symposium Proceedings* (1991).
2. C. Jackson, H. G. Barth, & W. W. Yau, "Polymer Characterization by SEC with Simultaneous Viscometry and Laser Light Scattering Measurements" *Waters International GPC Symposium Proceedings* (1991).
3. R. W. Klingler, *Starch*, **37**, 111-115 (1985).
4. M. A. Haney, D. Gillespie, W. W. Yau, *Today's Chemist at Work*, **3**, 11, 39-43 (1994)
5. Z. Grubisic, P. Rempp, and H. Benoit, *Poly. Lett.*, **5**, 753 (1967).
6. C. Jackson, H. G. Barth, & W. W. Yau, "Polymer Characterization by SEC with Simultaneous Viscometry and Laser Light Scattering Measurements" *Waters International GPC Symposium Proceedings* (1991).
7. Ptitsyn, O. B. Eizner, Yu E., *Sov. Phys. Tech. Phys.*, **4**, 1020 (1960).
8. Kratochvil, P. *Classical Light Scattering from Polymer Solutes*, Elsevier, Amsterdam (1987).
9. W. W. Yau, *Chemtracts-Macromolecular Chemistry*, **1**, 1-36 (1990).
10. W. W. Yau, *Chemtracts-Macromolecular Chemistry*, **1**, 1-36 (1990).
11. Z. Grubisic, P. Rempp, and H. Benoit, *Poly. Lett.*, **5**, 753 (1967).
12. D. J. Nagy, *J. Liquid. Chrom.*, **13**, 4, 677-691 (1990).
13. L. M. Gilbert, G. A. Gilbert, S. P. Sragg, *Meth. Carbodydr. Chem.* **4**, 25, (1964).
14. T. H. Mourey, S. M. Miller, W. T. Ferrar & T. R. Molaire, *Macromolecules*, **22**, 4286, (1989).
15. B. J. Goodfellow & R. H. Wilson, *Biopolymers*, **30**, 1183-1189 (1990).
16. A. Einstein, *Ann. Physik*, **19**, 289 (1906).
17. M. L. Fishman, D. T. Gillespie, & S. M. Sodney *Arch. of Biochem. and Biophys.*, **274**, 179-191 (1989).

17. D. E. Myslabodski, D. Stancioff, & R. A. Heckert, *Carbohydrate Polymers*, **31**, 83-92 (1996).
18. B. P. Wasserman & J. D. Timpa, "Rapid Quantitative Measurement of Extrusion-Induced Starch Fragmentation by Automated Gel Permeation Chromatography, *Starch* (1991).
19. M. S. Feather & J. F. Harris, *J. Am Chem. Soc.,* **89**, 5661 (1967).
20. I. Hall, D. Gillespie, K. Hammons, J. Li, *Adv. Chitin Sci,* **1**, 361-371 (1996).
21. G. G. S. Dutton & R. H. Walker, *Cell. Chem. Technol.,* **6**, 295 (1972)
22. B. H. Zimm, W. H. Stockmayer, *J. Chem. Phys.,* **17**, 12, (1949)
23. B. A. Lewis, M. J. S. Cyr & F. Smith, *Carbohydr. Res.*, **5**, 194 (1967)
24. L. A. Frannson in G. O. Aspinall, ed., *The Polysaccharides*, **3**, 337 (1985)
25. M. L. Fishman & P. D. Hoagland, *Carbohydrate Polymers*, **23**, 174-183 (1994).

Chapter 21

Application of Multi-Detector SEC with a Post Column Reaction System: Conformational Characterization of PGG-Glucans

Y. A. Guo[1], J. T. Park, A. S. Magee, and G. R. Ostroff

Alpha-Beta Technology, Inc., One Innovation Drive, Worcester, MA 01605

Multi-detector Size Exclusion Chromatography (SEC) with several post column reaction systems has been used to characterize the conformation of PGG-Glucans. PGG-Glucans are $\beta(1\rightarrow3)$ glucans isolated from the yeast cell wall (*saccharomyces cerevisiae*). Three conformational aspects of PGG-Glucans were characterized: 1) Single Chain (SC) conformation, 2) Triple Helical (TH) conformation, and 3) Triple Helical aggregates. PGG-Glucans were separated by SEC and analyzed using the refractive index, multi-angle laser light scattering, fluorescence, and polarimeter in combination with the post column reaction system. The Single Chain conformation was detected and characterized using SEC followed by Aniline Blue (AB) post column reaction system and using SEC coupled with polarimetric detection. The Triple Helical conformation was characterized using SEC coupled with polarimetry and sodium hydroxide post column reaction system. The Triple Helical aggregate was characterized using multi-detector SEC with and without sodium hydroxide post column reaction system. The Aggregate Number Distribution (AND) of PGG-Glucans across the entire molecular weight range was determined. The AND for Triple Helical PGG-Glucan ranged from 3 to over 10. These results indicate that PGG-Glucan forms aggregate of triple helical structure in aqueous solution.

PGG-Glucan, soluble $\beta(1\rightarrow6)$ branched $\beta(1\rightarrow3)$ glucan, is an immunomodulator that can enhance the host defenses by selectively priming neutrophil and monocyte/macrophage microbicidal activities without directly inducing leukocyte activation or stimulating the production of pro-inflammatory cytokines[1,2]. PGG-

---

[1]Current address: GelTex Pharmaceuticals, Inc., Nine Fourth Avenue, Waltham, MA 02154.

Glucan is isolated from the yeast cell wall of *saccharomyces cerevisiae*. The conformation of PGG-Glucan strongly affects its biological activities[2].

Deslande, Marchessault and Sarko[3] studied the conformation of curdlan, an unbranched ß(1→3) glucan, by X-ray diffraction and concluded that curdlan forms Triple Helical conformation. Thistlethwaite, Porter and Evans[4] studied Aniline Blue binding properties to ß(1→3) glucan and concluded that Aniline Blue binds to ß(1→3) glucan in NaOH aqueous solution. Our previous study showed that Aniline Blue binds specifically to Single Chain conformation[5]. Itou, Teramato, Matsuo and Suga[6] studied the optical rotation of Triple Helical schizophyllan and concluded that the specific rotation was +75 deg*cm^3 / g*dm in aqueous solution at the wavelength of 350 nm and 20°C. Hara, Kiho, Tanaka and Ukai[7] reported a value of the specific rotation of +19 deg*cm^3 / g*dm for Triple Helical ß(1→3) glucan at the wavelength of 586 nm and 20°C. Our previous study showed that PGG-Glucan gave negative optical rotation under alkaline condition[5].

Many other researchers also showed that ß(1→3) glucans form TH conformation in solutions[8-10]. We observed that the formation of TH was dependent on its single chain length. When the chain length is sufficiently long, polymer molecules are able to interact with each other via inter-chain hydrogen bonding, therefore, form TH or TH aggregate. However, when the polymer chain length is too short, it is incapable of forming strong inter-chain interactions, therefore, it remains in the SC conformation. The objective of this study was to develop methods to detect and characterize the SC, TH and TH aggregate conformations in soluble PGG-Glucans.

In this study, PGG-glucan conformers were separated in aqueous solution under pH 7 condition. The SC conformer was detected and characterized using a multi-detector SEC with a post column AB reaction system and a polarimeter. The TH conformer was characterized using SEC technique with a DRI and a polarimeter as the detectors. The aggregate state of TH conformation was determined through an Aggregate Number Distribution measurement using multi-detector SEC with a post column NaOH reaction system.

Experimental Section

Materials. Unfractionated soluble PGG-Glucan (Alpha-Beta Technology, Inc. Worcester, MA) was isolated from the yeast cell wall of *saccharomyces cerevisiae*. The purified TH and SC PGG-Glucan conformers were fractionated from the unfractionated soluble PGG-Glucan using a preparative SEC. Aniline Blue was purchased from Polyscience, Inc. Sodium nitrate (NaNO$_3$), HCl, and NaOH were purchased from EM Science.

Multi-detector SEC with a Post Column Reaction System. As shown in Figure 1, the multi-detector SEC with a post column reaction system consists of a pump (L-6000, Hitachi Instruments Inc.), an autosampler (AS-4000, Hitachi Instruments, Inc.), SEC columns (two KB804 and one KB803, Shodex), a post column mixing tee and a reaction coil (Upchurch). A post-column pump (L-6000, Hitachi Instrument Inc.) was

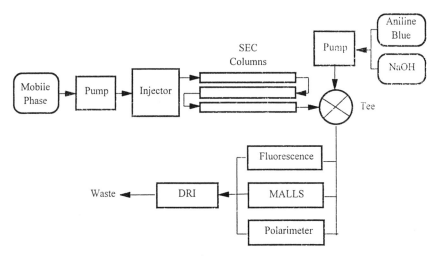

Figure 1. Block diagram of a multi-detector SEC with post-column reaction systems.

used to deliver the post-column reagents. A DRI detector (Sonntek) was used to determine the polymer concentration. A fluorescence detector (1046A, Hewlett Packard Co.) was used to detect Aniline Blue-SC complex. A polarimeter (AutoPol IV, Roudolph, Inc.) was used to measure the optical rotation of PGG-Glucans. A multi-angle laser light scattering detector (miniDAWN, Wyatt Technology Corp.) was used for the molecular weight determination. The mobile phase was 1.0 N $NaNO_3$. For SC detection, the post column mobile phase contained 1.0 mg/ml Aniline Blue. For TH aggregate number distribution determination, the post column mobile phase contained 866 mM NaOH. The flow rate was 0.5 ml/min for both mobile phase and the post column mobile phase. After mixing with $NaNO_3$, the final NaOH concentration that PGG-Glucans came in contact with was 433 mM. This condition is called pH 13 condition. The condition without post column reaction system is called pH 7 condition. pH 13 condition is fully disaggregated condition for PGG-Glucans and pH 7 condition is aggregate condition for PGG-Glucans[5].

Aniline Blue Binding. Commercial Aniline Blue was dissolved in water at 1 mg/mL. This Aniline Blue solution was activated by adjusting the pH to 12 for 1 hour using 1 N NaOH. Then, the activated Aniline Blue solution was neutralized to pH 7 using 1 N HCl. Upon activation and neutralization, much more Aniline Blue fluorophore was generated[5].

Molecular Weight and Molecular Weight Distribution. The molecular weight and the molecular weight distribution of PGG-Glucans were determined using the above laser light scattering detector and the above column separation system. An Astra® software version 4.2 was used for the molecular weight calculation. The refractive index increment (dn/dc) of PGG-Glucan was measured by Wyatt Technology, Corp., they are 0.143 ml/g and 0.145 ml/g at 633 nm under pH 7 and pH 13 conditions, respectively.

Aggregate Number Distribution (AND). PGG-Glucans were separated under pH 7 condition and detected under both pH 7 and pH 13 conditions. The aggregate number of PGG-Glucan was calculated using the pH 7 molecular weight divided by the pH 13 molecular weight for the same fraction in the DRI chromatogram. The AND is the distribution of the aggregate number across the entire molecular weight range of the SEC peak.

Results And Discussions

Detection of Single Chain PGG-Glucan Conformation. The Single Chain PGG-Glucan was detected using SEC with the post column Aniline Blue reaction system. The purified TH and SC PGG-Glucan conformers were injected into the SEC column and separated by size under pH 7. The Aniline Blue was delivered and mixed with the separated species through a post-column reaction system, and the fluorescence intensity was detected at the excitation and emission wavelength of 400 nm and 490 nm, respectively. Figure 2 shows the DRI and the fluorescence chromatograms. In

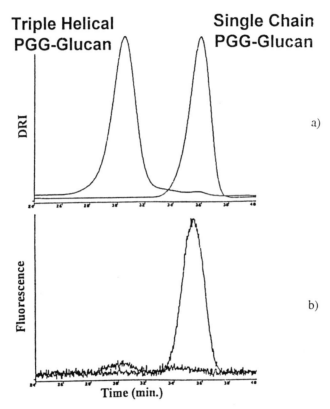

Figure 2. SEC chromatograms of SC and TH PGG-Glucan conformers obtained from a) DRI detector and, b) fluorescence detector with a post column Aniline Blue reaction system.

Figure 2, the fluorescence intensity of the SC PGG-Glucan is much higher than that for the TH PGG-Glucan, indicating that the SC PGG-Glucan forms a specific fluorescent complex with Aniline Blue.

Detection of the Ordered Triple Helical PGG-Glucan Conformation. The ordered conformation of TH PGG-Glucan and disordered conformation of SC PGG-Glucan were characterized using a multi-detector (polarimeter, DRI detector and multi-angle laser light scattering) SEC system. Figure 3 shows the DRI and the polarimetric chromatograms for the unfractionated soluble PGG-Glucan. In Figure 3, the SC PGG-Glucan fraction gave negative optical rotation. This is due to the chiral center of the ß(1→3) backbone linkage in the absence of physical aggregation. In contrast, the TH PGG-Glucan fraction gave positive optical rotation indicating an ordered conformation. The molecular weight distribution of unfractionated soluble PGG-Glucan ranged from 5,000 to 2 million g/mol. The results indicate that the conformational transition from SC to ordered TH conformation occurred at the molecular weight around 15,000 daltons.

Aggregate State of Triple Helical PGG-Glucan. The aggregate state of the TH PGG-Glucan was studied using a novel multi-detector SEC with a post column delivery system[11]. PGG-Glucans were separated under pH 7 or in the aggregated state. Sodium hydroxide was delivered after the column and mixed in-line with PGG-Glucan fractions to disaggregate the ordered PGG-Glucan conformer. The molecular weight was determined under pH 7 and pH 13 conditions. Figure 4 shows the molecular weight and molecular weight distribution for the TH and SC PGG-Glucan conformers. The SC conformation is confirmed by the similar value of the molecular weight under pH 7 and pH 13 conditions. This coincides with the results obtained from the Aniline Blue fluorescence and the polarimetry experiments. In contrast, TH PGG-Glucan showed evidence of ordered aggregation as indicated by the difference in the molecular weight under pH 7 and pH 13 conditions. The aggregate number for the TH and the SC PGG-Glucan conformers was calculated and presented in Figure 5. The aggregate number was determined to be one for the SC PGG-Glucan conformer and ranged from 3 to over 10 for the TH PGG-Glucan conformer and its aggregate. These results strongly indicate that PGG-Glucan isolated from the cell walls of yeast can form aggregate of triple helical structures. Many researchers reported that β-glucans (Scleroglucan, Lentinan, Schizophyllan) isolated from other sources form a single triple helix[9, 10].

Conclusions

1. PGG-Glucan can exist in single chain, triple helical and triple helical aggregate conformations depending on its single chain molecular weight. The aggregate number is one for the SC conformer and ranges from three to over ten for the TH or TH aggregate conformers. Schematic representation of the possible conformations of PGG-Glucan in aqueous solution is shown in Figure 6.
2. The TH conformer exists in an ordered conformation as indicated by the positive

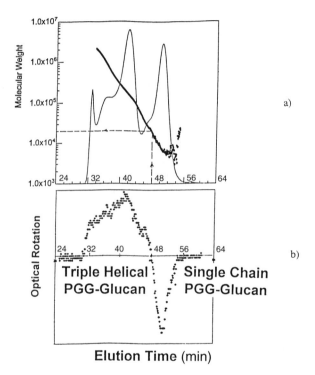

Figure 3.   Detection of SC and TH PGG-Glucan conformers in unfractionated PGG-Glucan using multi-detector (DRI, polarimetry, and MALLS) SEC system. a) The chromatogram was obtained from DRI detector and the molecular weight distribution was obtained from both DRI and MALLS detectors, b) The optical rotation chromatogram was obtained from a polarimetric detector.

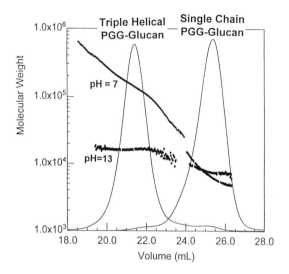

Figure 4. Molecular weight distribution of SC and TH PGG-Glucan conformers under pH 7 and pH 13 conditions, determined using multi-detector (DRI and MALLS) SEC with a post column NaOH reaction system for pH 13 condition and without a post-column reaction for pH 7 condition. The chromatograms were obtained from DRI detection.

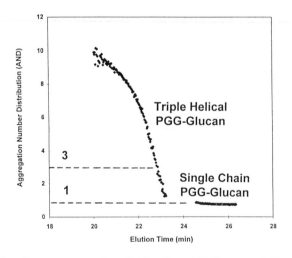

Figure 5. Plot of aggregate number distribution (AND) versus elution volume for the SC and TH PGG-Glucan conformers in neutral aqueous solution, determined by the multi-detector SEC with a post column NaOH reaction system.

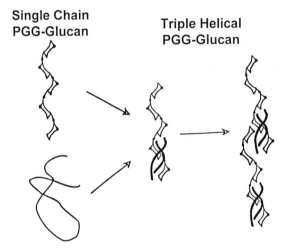

Figure 6.   Schematic representation of possible conformations of PGG-Glucans in aqueous solution, isolated from the yeast cell wall of *saccharomyces cerevisiae*.

optical rotation. In contrast, the optical rotation of the SC conformer is negative. The conformation transition occurred at the single chain molecular weight around 15,000 g/mol.

3. Multi-detector SEC with a post column reaction system is a powerful technique to study the aggregate number distribution of unfractionated PGG-Glucan across the entire molecular weight range.

Acknowledgments

Author acknowledge Roudolph, Inc. for lending the polarimeter model AutoPol IV.

References

1. Bleicher, P.; Mackin, W. *J. Biotechnology in Healthcare* **1995**, *2*, 207-222.

2. Jamas, S. B.; Easson, D. D., Jr.; Ostroff, G. R. and Onderdonk, A. B. *Polymeric Drugs and Delivery Systems;* ACS Proceedings; Edited by Dunn, R. L. and Ottenbrite, R. M., ACS: Washington D.C. **1991**, Serial No. 469, pp 44.

3. Deslandes, A. B.; Marchessault, C. D.; and Sarko. *Macromolecules* **1980**, *13 (6)*, 1466-1471.

4. Thistlethwaite, P.; Porter, I. and Evans, N. *Journal of Physical Chemistry* **1980**, *90(21)*, 5058-5063.

5. Park, J.T.; Guo Y.A. and Magee, A.S. *Proceedings of the ACS, Div. Carbohydrates* **1997**, San Francisco, April 13-17.

6. Itou, T.; Teramoto, A.; Matsuo, T. and Suga, H. *Macromolecules* **1986**, *19*, 1234-1240.

7. Hara, C.; Kiho, T.; Tanaka, Y; and Ukai, S., C. *Carbohydrate Research* **1982**, *110(1)*, 397-403.

8. Ogawa, K.; Tsurugi, J. and Watanabe, T. *Carbohydrate Research* **1973**, *29*, 397-403.

9. Yanaki, T.; Ito, W.; Tabata, K.; Kojima, T.; Norisuye, T.; Takano, N. and Fujita, H. *Biophysical Chemistry* **1983**, *17*, 337-342.

10. Norisuye, T.; Yanaki, T. and Fujita, H. *J. of Polymer Science: Polymer Physics Edition* **1980**, *18*, 547-558.

11. Park, J.T.; Guo, Y.A.; Vatsavayi, R. and Magee A.S. *Proceedings of the ACS, Div. Polymeric Materials: Science and Engineering* **1997**, 77, 58-59.

12. Guo, Y.A.; Park, J.T.; Vatsavayi, R.; Magee A.S. and Ostroff G.R. *Proceedings of the ACS, Div. Polymeric Materials: Science and Engineering* **1997**, 77, 60.

INDEXES

Author Index

Subject Index

Human immunoglobulin (IgG)
  assay method determining efficiency of
    antibody binding (a-IgG), 167–168
  coupling to poly(styrene/acrolein) parti-
    cles, 166
  derivatization of IgG, 165–166
  effect of attachment mode on anti-
    body-binding efficiency, 175
  efficiency of alkaline phosphatase con-
    jugate, 175, 176f
  experimental proteins for binding to la-
    tex particles, 165
  formation of antigen–antibody com-
    plex, 166–167
  thiolated IgG coupling to surfactant-
    coated polystyrene (PS) particles,
    166
  See also Colloidal particles; Sedimenta-
    tion field-flow fractionation (SdFFF)
Hyaluronic acid
  chromatogram, 305f
  molecular weight determination at col-
    umn exclusion limit, 304
  radius of gyration distribution, 306f
  See also Polysaccharide analysis by
    SEC³
Hydrodynamic volume, number average
  (H_n), experimental evaluation, 47
Hydroxyethyl cellulose (HEC)
  cellulose derivatives and structure, 117
  elution profile and differential distribu-
    tion by FFFF/MALLS/DRI, 133f
  See also Flow field-flow-fractionation
    (FFFF) with multi-angle laser light-
    scattering (MALLS) photometer and
    differential refractometry (DRI)
Hydroxyethyl starch (HES). See Starch
  derivatives

I

Immunodiagnostic latex particles
  types in model study, 163
  See also Colloidal particles
Immunoglobulin (IgG). See Human im-
  munoglobulin (IgG)
Immunology, use of colloidal particles,
  162–163
Interdetectors delay volume
  signal alignment for different detectors,
    73, 76
  See also Single-capillary viscometer
    (SCV) detector
Intrinsic viscosity (n)
  determination of Mark–Houwink–
    Sakurada parameters, 57, 58f

Mark–Houwink–Sakurada relationship,
    57
  See also SEC–viscometry
Ion exclusion
  disruption to ideal SEC separating
    mechanism, 121f
  size exclusion chromatography (SEC),
    120–122
  See also Fractionating methods
Isocratic elution with single-column
  packing
  coupling SEC with adsorption, 180
  coupling SEC with thermodynamic par-
    tition, 180–181
  liquid chromatography at critical ad-
    sorption point (LC CAP), 181–183
  liquid chromatography at theta exclu-
    sion–adsorption (LC TEA), 183
  See also Coupled liquid chromatogra-
    phy (LC) procedures
Iteration method, method for f_∞ and
  slope (K) determination, 10, 11f

K

Kinetics
  chain-end cleavage process, 267
  polymer degradation, 266–267
  random cleavage process, 266–267
  Simha–Montroll equation, 266

L

Latex particles
  experimental materials, 164–165
  human immunoglobulin (IgG) and
    anti-human IgG rabbit antibody pro-
    teins bound to surface of latex, 165
  types in model study, 163
  See also Colloidal particles; Sedimenta-
    tion field-flow fractionation (SdFFF)
LC PEAT. See Liquid chromatography at
  point of exclusion–adsorption transition (LC PEAT)
Lesec effect
  flow fluctuation phenomenon, 71
  precise flow in capillary, 234
  See also Single-capillary viscometer
    (SCV) detector
Light scattering, Raleigh equation, 252
Light-scattering detector
  branch analysis, 235
  combination of refractometer and
    PD2000, 237–238
  equations for light-scattering intensities
    of 90° and 15° angles, 237